SUCCESSION

# Advances in vegetation science  3

*Edited by*
EDDY VAN DER MAAREL

Dr. W. JUNK bv PUBLISHERS THE HAGUE – BOSTON – LONDON 1980

# Succession

*Symposium on advances
in vegetation sciences,
Nijmegen, The Netherlands, May 1979*

*Edited by*

EDDY VAN DER MAAREL

Dr. W. JUNK bv PUBLISHERS THE HAGUE – BOSTON – LONDON 1980

Distributors:
*for the United States and Canada*

Kluwer Boston, Inc.
190 Old Derby Street
Hingham, MA 02043
USA

*for all other countries*

Kluwer Academic Publishers Group
Distribution Center
P.O. Box 322
3300 AH Dordrecht
The Netherlands

ISBN-13: 978-94-009-9202-3          e-ISBN-13: 978-94-009-9200-9
DOI: 10.1007/978-94-009-9200-9

Reprinted from *Vegetatio,* Vol. 43, Nos. 1-2, 1980

# CONTENTS

# INTRODUCTION

Eddy VAN DER MAAREL

This volume is the second of two volumes covering the symposium 'Advances in vegetation science', which was held at Nijmegen, The Netherlands, from 15–19 May 1979. This symposium was organized on behalf of the Working Group for Data-Processing of the International Society for Vegetation Science. After this group held its final meeting two years earlier it decided to continue its activities, but in a wider scope. Most members of the Group felt that the original aim, i.e. the introduction of data-processing and multivariate methods for use in the systematic description of plant communities, was more or less fulfilled. The book Data-Processing in Phytosociology, largely based on papers in Vegetatio, edited by E. van der Maarel, L. Orlóci & S. Pignatti, and to be published by Dr. W. Junk Publishers, may be considered as a comprehensive report on the group's activities.

The wider scope I touched upon can be described as 'theoretical vegetation ecology' or 'descriptive and experimental analysis of vegetation'. Themes which would be appropriate within this scope include general models for vegetation analysis and for vegetation succession, general aspects of diversity and structural-functional properties. The further development of multivariate methods is certainly a matter of interest for the extended working group, but those methods should also be applied to other than synsystematical purposes, notably site-time series.

During the preparation of the symposium several colleagues showed their interest in showing results of applications of multivariate methods in succession studies. Hence it was decided to devote some sessions to succession problems, first of all to include these multivariate approaches, but also to pay attention to modelling and to the theory of vegetation dynamics.

As with the symposium theme Classification and ordination a very relevant manuscript was submitted at the time of the symposium withing being presented to the symposium itself. We considered it useful to include it in this proceedings volume.

All in all 16 contributions could be collected. The arrangement is as follows:

The contribution by Sjörs may serve as a general introduction to the types of changes and their names. The added paper by Noble and Slatyer provides appropriate facts and views on the mechanisms of vegetation dynamics.

Then a group of contributions follows in which data on species behaviour, plant demography and diversity during succession are discussed. This includes Faliński's study on sex structure and dynamics of pioneer woody species, a fynbos diversity study by Campbell & van der Meulen and studies on Mediterranean shrubs and trees in post-fire and postcultural developments by Trabaud and Lepart, Houssard, Escarré and Romane, and Debussche and Romane.

Fagerström and Ågren make a link with demography in their contribution on phenological spread.

A next group of papers by Austin, Persson, Regnell and van der Maarel deals with the multivariate approach to succession. The final group of contributions by Peet, Glenn-Lewin and van Hulst put succession again in a more general context.

Although this series of contributions is one out of rather many series and symposium volumes it may still find its own way, and direct further symposia on vegetation dynamics to be organized by our Working-Group, to begin with the Symposium in Montpellier, september 1980.

Finally I wish to remark that this symposium was sponsored by the Faculty of Science of the University of Nijmegen. I thank the Director of the Faculty, Dr. C.J.M. Aarts, for his support and cooperation and I am especially grateful for the considerable help which was given by Peter Toll, LL.D. of the Faculty's Directorate, both in the organization of the symposium and in taking care of the symposium guests! From our Department Rita Dubbers, Henk Butteling and Dr. Marinus Werger deserve being acknowledged for their help.

# AN ARRANGEMENT OF CHANGES ALONG GRADIENTS, WITH EXAMPLES FROM SUCCESSIONS IN BOREAL PEATLAND

Hugo SJÖRS

Institute of Ecological Botany, Uppsala University, Box 559–S-751 22 Uppsala, Sweden

Keywords:
Ecosystem change, Gradient, Peatland, Succession

## Omnipresence of change

Living nature is essentially dynamic. All biological systems operate on a flow basis, a flow of both matter and energy, a flow from source to sink. All biological systems are open and unstable, since life is a process and nothing would happen in life if living systems were completely closed and stable. This is true of all bio-dimensions, from the molecular to the global level.

It has often been argued that natural systems – especially those referred to as climaxes – tend to be self-stabilizing and that they represent steady states or even equilibria. However, even very old ecosystems such as coral reefs, tropical forests and savannas undergo short-term changes. There is no proof that they return to exactly the same composition in the long run, though their great number of cohabiting species may indicate a high degree of reproducibility over time and space. Other ecosystems, such as mires (peatlands), show long-term changes which in some cases seem to be unrelated to any climax at all.

## Natural changes

Ecosystem changes may be predominantly *biotically generated* or *environmentally conditioned*. The post-glacial invasion into Fennoscandia of the Norway spruce, *Picea abies*, is a good example of a biotically generated change affecting profoundly a large set of ecosystems. On the other hand, the impact of the crustal uplift producing new land in the same area is an equally good example of an environmentally conditioned change. Many changes are of both kinds, often in an intricate pattern of interactions. It may be stated that under

very adverse conditions of life, such as those prevailing in most of the Arctic or the higher alpine belts, successions are nearly always predominantly environmentally conditioned. But the same holds true also of some more favourable ecosystems such as dunes and deltas.

These changes are *allogenic*. It may be maintained that even the just mentioned invasion of the spruce, although biotically generated, is in principle allogenic. On the other extreme, some changes even on the ecosystem level are conditioned by such inherently intra-ecosystem events as growth and ageing of organisms, and should be regarded as *autogenic* in a restricted sense.

Another generally accepted classification of changes is in *fluctuations*, *cyclic developments*, and *trends*. In practice these categories may be very difficult to recognize and to keep separate.

This, I think, is one of the major difficulties in environmental as well as purely biological monitoring. To mention just one example, in the Kvikkjokk delta in Lappland, one five-year period was characterized by frequent high floods and great sedimentation almost every year, but another five-year period by the opposite conditions. The corresponding ecosystems changed greatly in the period of many great floods, but developed slowly towards stabilization during the nearly flood-less sequence of years.

A slightly different subdivision of changes is in *non-persistent* and *persistent*. Only the fairly persistent changes are usually included in the term successions.

Successions may be *reversible* but *irreversible* successions are more general.

Some successions are regarded as *progressive*, others as *retrogressive*. It is exceedingly difficult to define these concepts, except in extreme cases. Good examples are given in Eva Waldemarson Jensén (1979). Increasing complexity, for instance addition of a forest layer to a previously treeless type of vegetation, is obviously progressive, as is in most cases increasing productivity. Retrogressive, on the other hand, refers to returning to a former, less well-developed state, or to destruction of already built-up structures. Successions leading to, or caused by, increasing wetness are often retrogressive, for example, the formation of bog hollows or of the structures known as *flarks* or *rimpis* on the boreal aapa mires.

There are other possible subdivisions of successions, for instance that in *primary* successions starting more or less from scratch, and *secondary* ones involving shifts in developmental directions in already developed ecosystems, obviously the majority of cases. However, the definitions of primary and secondary successions differ considerably between authors.

Some sets of successions increase the similarity between the ecosystems concerned and may be regarded as *convergent*, whereas other sets of successions lead to increasing difference and could therefore be termed *divergent*. The creation of hummocks, hollows and pools on a bog surface is an example of a divergent set of successions. By the way, the existence of divergent trends in the successions within an area is proof of the invalidity of the doctrine of mono-climax.

Changes can also be characterized by their *duration*. There is a time-scale from the day-and-night and the annual happenings to the century-long rotation period in the forest and even to changes over the millennia, not to speak of the hundreds of thousands of years needed for major geological changes and biological evolution.

### Man-made changes

To what extent do man-made changes differ from those conditioned by nature alone? I would say, not very much in principle but quite a lot in performance.

Man-made changes may be fluctuations or trends but are less frequently cyclic occurrences. They may cause persistent successions or be non-persistent, or just cause marginal effects in the functioning of the ecosystem without altering it profoundly. The more important man-made changes usually have a duration of decades or centuries, though in countries with ancient civilizations like those of the East and the Mediterranean major man-made ecological changes date back between five and nine thousand years.

Table 1. 'Ordination' of ecosystem changes.

Biotically generated – environmentally conditioned changes
Autogenic – allogenic successions
Fluctuations – cyclic changes – trends
Non-persistent changes – persistent successions
Reversible – irreversible successions
Progressive – retrogressive successions
Primary – secondary successions
Convergent – divergent sets of successions
Short-term – medium-term – long-term successions

Natural – man-made changes:
completely unintentional
intentional but with unexpected consequences
intentional, with foreseen but unwanted consequences
intentional, with planned, wanted consequences

Ecosystems influenced by man, and thus changed, are often looked upon as less stable than natural ones. Obviously the human impact *per se* adds change to an even originally changeable world. But provided the human impact remains reasonably constant a fairly stable ecosystem may result, such as for instance a meadow used for hay-making every year.

I think it is important to put the emphasis on the kind of action by man. He may work by changing either the biotic part of an ecosystem or the non-living environment. Frequently, however, man may simultaneously change both the environment and the biotic community.

There is one unique feature in the man-made changes, as distinguished from natural ones, and this is *intention*. Most of man's actions are intentional, but their consequences may not be so. When we discuss the impact of man, we therefore have to think of his aims and if he *is aware or not of the consequences* of his actions, and finally in case he is aware, if the consequences are *unwanted* or wanted, that is *planned*.

Much of the really bad impact of man into his own and other living beings' environment was made by *deliberate* neglect of the consequences of actions which were made voluntarily, but there are also cases when environmental disturbance was made in good faith.

We anticipate a much more profound human impact not only on urban or industrial areas but even on rural ecosystems in the near future. For instance, forests and wetlands will be changed by human activities even in regions only moderately affected up to the present time.

2

## Origin of peatland

Ecological textbooks often exemplify progressive succession with the hydrarch development or *hydrosere*, starting from shallow water and leading to fen and eventually to bog or even wooded bog. However this is not the only way of peatland formation, and not even the dominant kind of origin. The least discussed origin is the *primary mire formation* which is a primary succession starting directly upon moist soil exposed to colonization either after glacier retreat, or by crustal uplift from the sea (in places also from freshwater). Other situations where peat may be formed by way of primary successions are in some dune slacks, on some fresh delta deposits, and on some other flatlands formed by shore processes.

By far more common is the origin of peatland from secondary succession by way of so-called *paludification*. This development may perhaps be termed retrogressive, especially when the original vegetation was a forest.

The origin of most of the world's peatlands is still unknown, and maybe the frequency of direct, primary peat formation just mentioned has been underestimated. However, it is pretty certain that nearly all sloping peatlands (e.g. blanket bogs, sloping fens and most of the aapa mires) were formed by paludification in post-glacial time. Very likely also much of the extremely extensive peatlands in the plains of the north and interior of Eurasia and North America were once covered by woodland or in some cases grassland.

## Later successions on mires

The present vegetation and structure of peatlands are not clearly related to their origin, and in some cases even nearly independent of the way of origin. In consequence, the general trend of peatland successions may be described as *convergent* in a very broad, over-all sense. But this does not lead to anything like conformity. Successions in peatland are still to a very great extent dependent on external factors, that is, environmentally conditioned, besides being, of course, also biotically generated. The extent of environmental conditioning generally increases towards north. Also ruggedness of the landscape, volcanism, and of course human influence, contribute to the environmentally conditioned diversity.

But even the biotically generated, more or less autogenic development does not always lead to uniformity. It is true that some peatlands like the one shown by

Clymo (1980) may be spatially uniform. But most raised bogs and the aapa mires of the north develop surface patterns consisting of drier and wetter parts, which did not exist in the early stages of development and in consequence are the results of divergent sets of secondary successions. The fact that nearly all such patterns show an orientation at right angles to the slope is proof that they are not exclusively phenomena of biotic succession, but rather produced by an interaction between biotic and environmental factors within a gradually diversified ecosystem. There are many striking examples of patterned peatland showing various kinds of divergence and retrogression.

Some of these successions were extensively dealt with by mire scientists of the past generation, and some were thought of as cyclic successions, others as unidirectional trends. However, the extent to which truly cyclic successions occur on the raised bogs was evidently over-estimated, and more often the changes must be regarded as persistent. This is true of such striking features as the high strings on northern raised bogs known in Finnish as *kermis*, and the large wet *algal hollows* often developed on bogs. Other permanent features developed in the past by divergent sets of partly retrogressive, irreversible, secondary successions are the *bog-pools* of many bogs, especially in central Fennoscandia but also found elsewhere. The same is true of the much-discussed *string-and-flark patterns* of the northern aapa-mires. However, as we move northward, the patterns become increasingly dependent on two sets of external factors, namely, first, hydrology of the peatlands as constituting outflow areas for both meltwater and groundwater of the whole landscape and, secondly, frost action of various types. The permafrosted *palsas* of northernmost Fennoscandia and other sub-arctic areas are extreme examples of features created nearly exclusively by environmentally conditioned succession, since the biotic components of these successions are confined to the dying-off of the peatforming vegetation and its replacement by a thin cover of species which can grow under the stress of high acidity, frequent surface dryness and even erosion by wind and rain.

## Time dimensions

The rate of peatland processes is usually slow. Some processes may be observable in a human life-time, but in a case investigated (Backéus 1972) there were few changes on the surface of a large raised bog within 60 years. In areas where peat had been excavated some quite strong successions have been demonstrated to occur within a few centuries, but the major geological events have a time-scale of millennia. However, within the geological record, some phases are less slow than others. Especially the very early postglacial and the two last millennia B.C. seem to have been periods of comparatively rapid and profound changes, whereas our own time is one of relatively great stability as far as natural changes are concerned.

Unfortunately, the same cannot be said about man-made changes. To an extent which gives us great concern, the peat-forming ecosystems are now being destroyed or even consumed, and peatland conservation should be given high priority in all countries.

## An analogy to ordination

By analogy, the present paper was presented as an 'ordination', within quotation marks, although completely non-numerical. This was because a multidimensional arrangement along directions of variation or gradients was attempted, referring to the nature of changes in ecosystems. This arrangement can be looked upon in a similar way as you look upon the axes found when an ordination procedure has been carried out. In fact, the multidimensional gradient approach preceded ordination in the development of vegetation science and is the logical background of ordination techniques.

## Summary

A multidimensional arrangement of various gradients referring to the nature of ecosystem changes is presented (Table 1). Examples are given, some of them from boreal peatlands (mires) which show both convergence from different origin and divergence in the local development. Some successions in wetlands are retrogressive, and the changes are usually at the same time biotically generated and environmentally conditioned, sometimes (e.g. shores, deltas, palsas) mainly the latter.

## References

Clymo, R.S. Preliminary survey of the peat-bog Hummell Knowe Moss using various numerical methods. In: E. van der Maarel (ed) Advances in vegetation science: Classification and ordination. Vegetatio 42:. . . –. . . .
Waldemarson Jensén, E. 1979. Successions in relationship to lagoon development in the Laitaure Delta, N Sweden. Acta Phytogeogr. Suec. 66 Uppsala

Accepted 6 November 1979

# THE USE OF VITAL ATTRIBUTES TO PREDICT SUCCESSIONAL CHANGES IN PLANT COMMUNITIES SUBJECT TO RECURRENT DISTURBANCES

I.R. NOBLE & R.O. SLATYER*

Department of Environmental Biology, Research School of Biological Sciences, PO Box 475, Canberra City, ACT 2601, Australia

**Keywords:**
Disturbance, Dynamics, Fire, Models, Succession, Vegetation

## Introduction

The established view of ecological succession is that, following a disturbance, several assemblages of species progressively occupy a site, each giving way to its successor until a community finally develops which is able to reproduce itself indefinitely. Implicit in this view is the assumption that each suite of species modifies the site conditions so that they become less suitable for its own persistence and more suitable for its successor, and the assumption that only the final community is at equilibrium with the prevailing environment. These ideas owe their origin largely to Clements (1916, 1936) who viewed the community as a kind of super-organism, and succession as a form of ontogeny. They are entrenched to various degrees in the ecological literature and have been supported by many authors (see, for example, Golley 1977).

It is a matter of common observation that, with the passage of time following a disturbance, shifts in dominance may occur with ephemeral herbaceous life-forms progressively becoming overtopped by taller perennial herbs, shrubs and trees. However these shifts may not reflect the progressive entry to the community of the taller long-lived forms, but instead, the gradual emergence and dominance of species which may have been present, but inconspicuous directly after the disturbance.

In a perceptive discussion of the principles involved in vegetation development, Egler (1954) concluded that in many situations the 'initial floristic composition' following

disturbance determines the future shifts in dominance, various species successively becoming dominant as their life history characteristics, and associated life-forms, are exhibited with the passage of time. Egler considered that classical successional patterns, 'relay floristics', may be much less widespread than normally assumed and may be associated mainly with the entry of species to the community long after the original disturbance. The importance of site occupancy by a particular species, or suite of species, was emphasized by Egler as a factor restricting the subsequent entry of other species.

In drawing together the work of Egler, and other workers who have emphasized the failure of classical succession concepts to describe vegetation development in many situations, Connell & Slatyer (1977) have suggested recently that most successional sequences involve one of three main types of pathway. The first, '*facilitation pathway*', is essentially the classical, relay floristic pathway, in which the presence of early occupants facilitates the entry of successive suites of species. The second, '*tolerance pathway*', describes the situation in which later species are successful whether or not earlier species have preeded them; they can become established and grow to maturity in the presence of other species because they can grow at lower levels of resources. The third '*inhibition pathway*', describes the situation in which later species cannot grow to maturity in the presence of earlier ones. Unless they are present on the site their entry may be inhibited by the early occupants, thereby leading to dominance by species not normally regarded as late succession species.

Studies such as Connell and Slatyer's (for example see also Horn 1976) are based on community properties

* We wish to thank A.M. Gill, A.N. Gillison and B.R. Trenbath for their critical reading of draft manuscripts, and P.M. Cochrane for assistance with many aspects of manuscript preparation.

but implicitly emphasize the importance of the life history characteristics of individual species in determining the patterns of succession. The importance of individual properties was explicitly recognized by Gleason (1926) many years ago and recently emphasized by Drury & Nisbet (1973) who concluded that most of the phenomena of succession could be understood as consequences of differential colonizing ability, growth and survival of species adapted to grow in different environments.

Given the relative availability of species at and near a disturbed site, it is not surprising that the observed replacement sequences tend to be reproducable. The observed patterns need not reflect a form of biological determination of the type envisaged by Clements, but rather an inevitable consequence of the relative availability of a range of species and their life history characteristics.

These considerations, and the mass of literature concerning succession, enable several generalizations to be made about major factors affecting succession. These are:
1. Species composition immediately after a disturbance is dependent upon propagules, which have either dispersed from elsewhere or have persisted through the disturbance at the site, or upon vegetative resprouting from organs surviving the disturbance. (Our use of the term 'propagule' is defined in Appendix 1).
2. Immediately after a disturbance there is a pulse of recruitment or regrowth under conditions of little competition for space or other resources.
3. After the initial pulse, recruitment slows since once an individual plant is established it is very difficult to displace.
4. Subsequent recruitment of additional species is sometimes facilitated by prior occupancy, but is frequently restricted and may be inhibited.
5. In the absence of further disturbances, long lived species and those which can regenerate in the presence of their own adults will finally become dominant.

Using these generalizations, Noble & Slatyer (1977, 1978) developed a scheme for predicting the major shifts in species composition and dominance in plant communities subject to recurrent disturbance. Early versions of the scheme have been applied with satisfactory results to such diverse situations as fire in temperate rainforest, shrub and woodland communities in Australia (Noble & Slatyer 1977, 1978) and mixed coniferous forests in western North America (Cattelino et al. 1979).

The scheme is intended to deal mainly with succession in a community at a particular site with a stable physiography and in the absence of climatic change. A site is assumed to contain a representative sample of the community. The scheme has been derived to deal mainly with terrestrial communities dominated by higher plants, although it is probably applicable to a wider range of situations. The scheme utilizes a small number of life history characteristics pertaining to the potentially dominant species in a particular community. These characteristics are termed 'vital attributes' since they are vital to the role of the species in a vegetation replacement sequence.

In the present paper a rigorous evaluation of the scheme is presented. First, we provide a detailed account of the derivation of the vital attributes, leading to a set of the most common combinations of attributes. We then provide examples of replacement sequences for wet sclerophyll and rainforest communities in south-eastern Australia and a mixed coniferous forest community in north-western U.S.A. Factors which cause variations in vital attributes are then examined, together with the effect of various types of disturbances on patterns of replacement sequences.

## Vital attributes

Vital attributes are those attributes of a species which are vital to its role in a vegetation replacement sequence. We see the following groups of vital attributes as the most important:
1. The method of arrival or persistence of the species at the site during and after a disturbance.
2. The ability to establish and grow to maturity in the developing community.
3. The time taken for the species to reach critical life stages. More than one biological mechanism or phenomenon may be responsible for a particular vital attribute displayed by a species, and for a given vital attribute the biological mechanisms may differ from species to species. The vital attribute reflects only the outcome of these mechanisms.

### Definition of life stages

Before discussing the vital attributes in detail, certain important terms describing the life stages of a species need to be defined. Four major stages are recognised, each of which applies to the population of the species at the site, and not to particular individuals.
1. At the *juvenile* stage the majority of individuals in the population are immature. Juvenile individuals may have

methods of persisting through a disturbance by vegetative means.

2. At the *mature* stage, many (or all) of the individuals in the population are reproductively mature and able to produce propagules. They may also be able to persist by vegetative mechanisms.

3. If only propagules remain available, arising either from the original population and stored at the site, or from dispersal from surrounding populations, then the species is said to be at the *propagule* stage. At this stage there no longer persists a significant number of juveniles or adults at the site.

4. If, following the loss of the original adult population either through senescence or because of a disturbance, a stage is reached where no propagules of the species are available at the site, the species is completely lost from the site and is said to be *locally extinct*.

The boundaries between the above four life stages of a population are not perfectly distinct, since transitions do not occur at an instant in time, and in some cases the onset of different mechanisms of persistence may not exactly coincide. However satisfactory empirical rules can be found to handle such situations.

### Persistence and arrival

The first group of vital attributes deals with the method of persistence through the disturbance. The distinction between asexual and sexual forms of persistence has been described by many authors (Jarrett & Petrie 1929, Naveh 1975, Gill 1975, Lyón & Stickney 1977). Although this is a fundamental division within the methods of persistence, in dealing with vital attributes we wish to emphasize not so much the mechanism, but the outcome of the mechanisms possessed by a species.

A mechanism leading to persistence through a disturbance may not be operative during all life stages of a species. For example, a species which has a seed store in the canopy as its sole mechanism of persistence, will have this mechanism available only during the mature life stage. If we consider the four life stages recognised above, and the fundamental difference between sexual and asexual forms of reproduction, 20 patterns of availability are theoretically possible (Fig. 1). Since vital attributes refer to the outcome of various mechanisms, 20 different types of vital attribute associated with persistence and arrival are possible. However, we have concluded that only 10 of these patterns are biologically

feasible, and only 8 are common in nature, using the following arguments.

*Propagule based mechanisms*

The first seven patterns in Fig. 1 are characterized by propagule availability for an indefinitely long period after

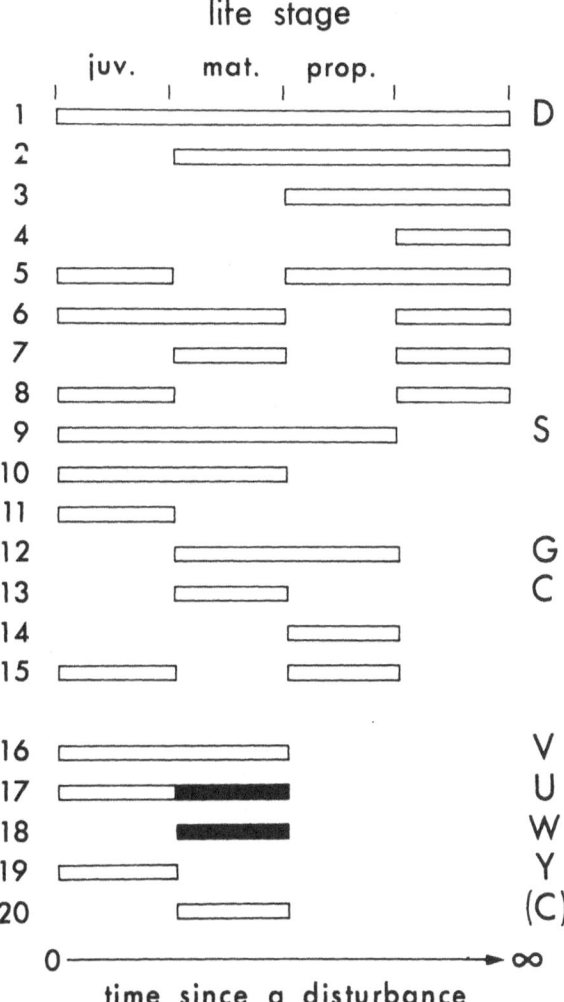

Fig. 1. Diagrammatic representation of patterns of availability of a method of persistence in relation to the four life stages. Patterns 1 to 15 show each of the possible patterns associated with a propagule based mechanism, and patterns 16 to 20 those arising from a vegetatively based mechanism. An open bar indicates that a method of persistence is available at a particular life stage of a species population. The method will usually result in only juvenile material being present immediately after a disturbance, but in some cases (solid bar) mature tissue will persist. The symbol for the vital attribute associated with those patterns judged to be biologically feasible is shown on the right.

the loss of mature individuals from the site. Consequently, these patterns require propagule storage beyond the boundaries of the disturbed area and involve dispersal mechanisms to enable the propagules to reach the site.

Pattern 1 is very common in nature, being found in species with propagules which are dispersed over an area large enough for them to be available at any time that a disturbance occurs in a part of that area (i.e. the mechanism is available at any life stage of the species). This pattern will apply mainly to seed dispersal, but some vegetative propagules are capable of long distance dispersal, particularly by water or animals.

Patterns 2–8 also require a mechanism which is available for an indefinitely long time after the loss of mature individuals of a species from the site; a requirement which can only be met by dispersal from elsewhere. However, all these patterns include a life stage during which the mechanism is not available. This means that if a site is disturbed while the species is in that particular stage, then any propagules dispersing to the site, both immediately after the disturbance or any time thereafter, will not be able to establish. We see no biologically feasible mechanism which leads to such patterns and consequently it seems unlikely that patterns 2–8 will be found in nature.

The next group, patterns 9–15, include the main mechanisms by which propagules can persist through a disturbance at the site.

Pattern 9, in which propagules are available at all life stages extending beyond the death of mature individuals until the local extinction of the seed pool, is common in nature. It is well represented in species with seeds capable of maintaining viability in the soil, or in protective fruits, for periods which may be substantially longer than the life span of the individuals.

Pattern 10 appears to be unlikely. It implies that the mechanism is lost as soon as the adults disappear from the site through senescence and death. However, if the site is disturbed at any stage before this event the mechanism operates, not only at the first disturbance, but also if a subsequent disturbance occurs before the propagule pool is replenished by maturing individuals. It seems contradictory that the propagules do not maintain viability if the adults die after senescing, but do retain viability if the adults die due to a disturbance. It is more likely that the propagules retain some viability after senescence, leading to pattern 9; or the propagules are viable only while adults are present, leading to pattern 13.

Pattern 11 seems to be very unlikely. If it were to

operate via seed produced at the site, the pattern implies that the mechanism would no longer operate after the species matured at the site. The maturation of the species could only be prevented by frequent disturbances, but in this case the seed pool would never be replenished and would be depleted rapidly. A pattern of this form may be possible via a vegetative mechanism of persistence, and is discussed as pattern 19.

Pattern 12 is similar to pattern 9 except that the persistence mechanism is not available if the disturbance occurs while the species is in the juvenile stage. This pattern will occur if, for example, a long lived seed pool is exhausted by germinations and seed deaths after the first disturbance. Until the species becomes reproductively mature there will be no seed pool to allow re-establishment after a subsequent disturbance.

Pattern 13 is very common occurring, for example, in species which retain a short lived seed pool which can survive and germinate immediately after a disturbance. Unlike pattern 9, the seed pool is so short lived that there is effectively no survival beyond the adult stage.

Pattern 14 implies that none of the propagules become viable until after the death of the adults due to causes other than disturbance. This seems to be a most unlikely pattern. It should be noted that monocarpic species, which fruit only once and then die, do not show pattern 14, but rather the mature stage is reduced to a single point and pattern 12 provides a more correct description.

Pattern 15 is possible but probably rare. The seed pool becomes non-viable at the mature stage, but becomes viable again when the species senesces and dies. One possible mechanism would involve a species which affected the intensity of the disturbance (e.g. by providing fuel for fires) in such a way that the disturbance is sufficiently intense during the mature stage to destroy the seed pool, but is not so during other stages. It is also possible, but even less likely, that such a mechanism could lead to pattern 14 if the juvenile stage also affects the intensity of the disturbance so as to destroy the seed pool.

In summary, of the 15 propagule based mechanisms we see only four as being commonly found in nature; these are patterns 1, 9, 12 and 13. Two other patterns (14 and 15) are biologically feasible but probably very rare. Bizarre mechanisms could probably be proposed for some of the other patterns, but the essential point remains that only four propagule based patterns are commonly found in nature.

Throughout this paper we will refer to vital attributes

8

derived from these four patterns of mechanism availability as follows: pattern 1 as D for *dispersed* propagules; pattern 9 as S for long lived propagule *store*; its variant (pattern 12) as G since the whole propagule store either *germinates* or is otherwise lost at the first disturbance, and pattern 13 as C since a common mechanism involves the storage of short lived propagules in the *canopy* of the individuals.

*Vegetative based mechanisms*

Turning to the five vegetative-based mechanisms it is evident that pattern 16 is feasible and common. Species which have the ability to root sprout or to sprout from a burl or lignotuber show this pattern. The essential feature of the mechanism is that the individuals survive the disturbance, but lose all of their reproductively mature tissue, and must regrow for at least a short period before becoming mature again.

Pattern 17 is a variant of the previous pattern in which, if the species was present at the adult stage before the disturbance, then it remains reproductively mature after the disturbance. Species which are able to survive a disturbance with little harm, or species which regrow very rapidly will show this pattern.

Pattern 18 occurs if the adults are able to persist through the disturbance as above, but where the juveniles cannot do so. An example is a species which has thick barked adults which are resistant to fire, but which has juveniles which are thin barked (and shorter) and which succumb to the fire.

Pattern 19 appears to be biologically feasible although only after certain types of disturbance. It will occur if the juveniles of the species are able to resist the disturbance, but the adults cannot do so. Such a mechanism may be possible in disturbances in the form of high wind, floods or possibly some pest epidemics, but would be most unlikely if the disturbance is fire.

Pattern 20 occurs where only the adults can survive a disturbance but they lose all of their reproductively mature tissue. This pattern is identical in outcome to pattern 13, even though the biological mechanism is different. Since we are interested not so much in the biological mechanism, but rather, in the outcome of it, species with this mechanism can be regarded as having the C vital attribute.

Throughout this paper we will refer to the vital attributes derived from these patterns as follows: pattern 16 as V for *vegetative*; pattern 17 as U since the species is virtually *unaffected* by the disturbance, and pattern 18 is W (hence the three major vegetative patterns are U, V and W). Pattern 19 will be included in the next section, which considers combinations of vital attributes. However, we regard the possibility of a species showing a mechanism of persistence leading to this pattern alone as unlikely, and will not include the pattern as a vital attribute. The symbol Y (*young*) will be used for this pattern.

*Combinations and hierarchies of mechanisms*

In examining all the possible patterns we have concluded that eight are biologically feasible and probably common in nature (i.e. those associated with the vital attributes D, S, G, C and V, U, W, Y). It is also possible for a species to have a combination of two or more mechanisms of persistence. For example a species may be thick barked and able to resist a fire at the adult stage, and therefore would be said to have the W vital attribute if this was its only mechanism of persistence. But it may also have well dispersed propagules and therefore would be said to have the D vital attribute if it had this mechanism alone. The combination of mechanisms forms a pattern of availability not encountered before and species displaying this pattern are defined to have vital attribute $\varDelta$ (Table 1). Another example is a species which is able to resprout from underground after a disturbance which kills its top (V) and which also has a long lived seed store in the soil (S). In this case the combination of these patterns of availability is still equivalent to that of S alone, although the specific mechanism whereby the species persists may be a mixture of both the propagule based and the vegetative mechanisms.

The complete set of combinations of pairs of mechanisms with different patterns of availability is shown in Table 1. Only three new, derived patterns of availability occur, and each is feasible. The vital attributes associated with these derived patterns have been labelled with the Greek characters $\varDelta$, $\varSigma$ and $\varGamma$ since they are very similar to the D, S and G vital attributes except at the adult stage. Combinations of three, or more, mechanisms lead to no new patterns.

Table 1. Vital attributes assigned when a species possesses two mechanisms for persistence.

|   | D | S | G | C | V | U | W |
|---|---|---|---|---|---|---|---|
| S | D+S=D |  |  |  |  |  |  |
| G | D+G=D | S+G=S |  |  |  |  |  |
| C | D+C=D | S+C=S | G+C=G |  |  |  |  |
| V | D+V=D | S+V=S | G+V=S | C+V=V |  |  |  |
| U | D+U=Δ | S+U=Σ | G+U=Σ | C+U=U | V+U=U |  |  |
| W | D+W=Δ | S+W=Σ | G+W=Γ | C+W=W | V+W=U | U+W=U |  |
| Y | D+Y=D | S+Y=S | G+Y=S | C+Y=V | V+Y=V | U+Y=U | W+Y=U |

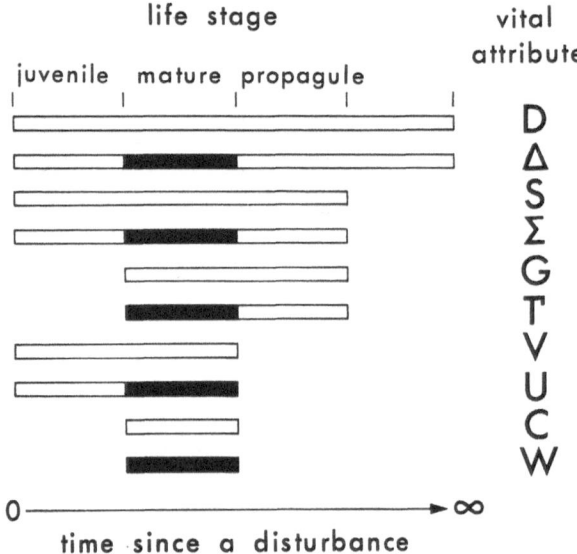

Fig. 2. A summary of the ten vital attributes associated with the method of persistence. The symbols used to represent the vital attributes are shown on the right. The open and solid bars have the same meaning as in Fig. 1.

The complete analysis of the combinations of mechanisms with different patterns show that ten distinct patterns are likely to be observed in nature (Fig. 2). This means that a species can display one of ten different vital attributes describing the method of persistence. It should be noted that the patterns are of the form of D, S, G, V and C, with and without a vegetative mechanism of persistence at the adult stage. It is, of course, possible that a species has no mechanism of persistence for a given disturbance type, however it would become locally extinct at the first disturbance and is therefore of little interest in this scheme.

**Establishment and maturation**

The second group of vital attributes describes the conditions under which a species which has propagules (or juvenile vegetative resprouts) available at a site can establish and grow to maturity. Again we are interested not so much in the precise conditions required for ecesis but rather when, in relation to a disturbance, the species can establish and grow to maturity. As discussed above, there are two stages in community development after a disturbance. The first, immediately after a disturbance and during the period when there is little competition for resources,

Fig. 3. The four patterns and the three vital attributes associated with the conditions required for establishment. The dashed section of the bars indicates that the I and R mechanisms are not necessarily synchronous.

and the second, after this initial pulse, when competition for resources becomes progressively more important. There are four patterns of availability related to these two stages, which are depicted in Fig. 3.

Pattern 1 is found in species which are able to establish and grow at a site immediately after a disturbance, and continue establishing and growing in the conditions of increased competition. This results in a mixed age structure even though recruitment may be most rapid immediately after a disturbance. An age structure of this type may develop when recruitment occurs singly, following the death of an established individual. Such gap-phase replacement could be thought of as indicating pattern 2, if the site under consideration was as small as a single individual. However, since we are dealing with representative samples of whole communities gap-phase replacement will lead to the mixed age structure characteristic of pattern 1.

Pattern 2 is demonstrated by species which are able to establish and grow at a site immediately after a disturbance, but cannot continue recruitment as the competition for resources increases. Many species, including 'pioneer' species, show this pattern which results in uniform-age cohorts which correspond to the dates of previous disturbances. If the species population is killed by each disturbance only a single age cohort will be found.

Pattern 3 includes those species which are unable to establish immediately after the disturbance, but require some conditions associated with an older community for recruitment. This may be associated with a delay so that either inhibitory compounds can be leached, or so that dispersal of propagules might occur (as in the case of large fruited species which rely on animal dispersal). This pattern results in mixed age cohorts, the oldest of which is younger than the time since the last disturbance.

Pattern 4 suggests a mechanism which is not available

immediately after a disturbance or in a long-undisturbed community, but only at an intermediate stage. It could conceiveably be found in plant species which require some conditions associated with an older community – such as dispersal or site preparation by animals – which do not continue to prevail with the passage of time; for example if a key habitat requirement for the animals eventually is lost from the community. However, we have not found any examples of such species and have concluded that this pattern, if it occurs at all, is rare.

In summary, of the four patterns of availability of a mechanism for establishment and growth to maturity, only three appear to be common in nature. We refer to species exhibiting pattern 1 as T species, because they *tolerate* a wide range of site conditions, pattern 2 as I species because they are *intolerant* of competition, and

pattern 2 as R species, because they *require* some condition present in established communities.

We are conscious that there may be subtle mechanisms leading to vital attributes which are not reflected in Figs. 2 and 3. Should other patterns exist they can be readily included.

## Species types

We consider these first two vital attribute groups as the most fundamental, and a description of a species in terms of these two attribute groups will tell much about its role in vegetation replacement sequences. Since there are ten categories of the first vital attribute group and three of the second, 30 different species types are possible. How-

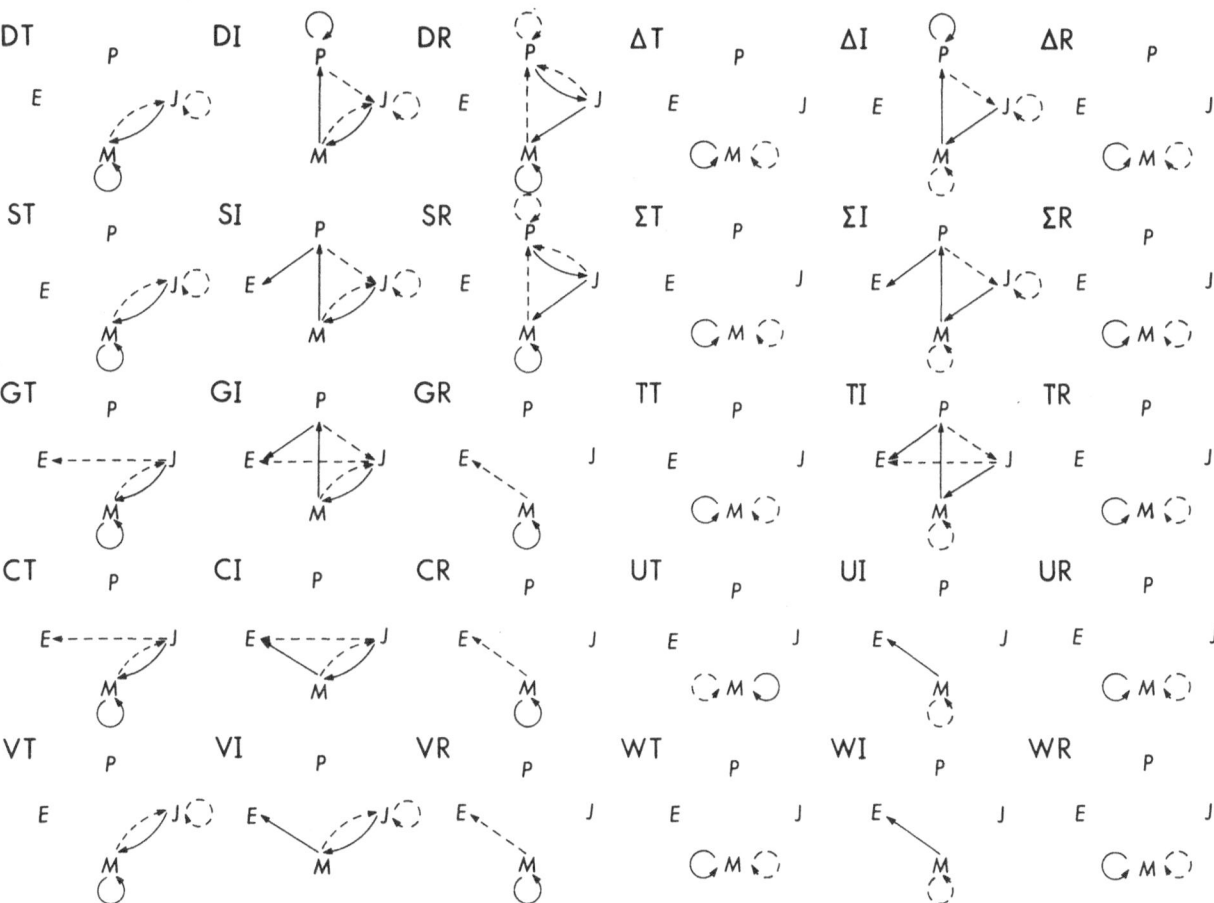

Fig. 4. Transitions between life stages for each species type. Solid arrows show the transitions in periods with no disturbance, and broken line arrows show the transitions due to a disturbance. The letters J, M, P and E refer to juveniles, mature, propagule and extinct, life stages, respectively.

ever, not all of the different species types show unique behaviour in replacement sequences. This is most clearly demonstrated by drawing 'transition diagrams' for each species description.

### Transition diagrams

The transition diagrams show the sequence of life stages that a species will pass through as a result of a particular disturbance regime. The transition diagram deals with the four major life stages of a species, i.e. juvenile (J), mature (M), propagule (P), and locally extinct (E).

The logical starting point in a transition diagram is the propagule stage, representing the initial invasion of a site. However it is convenient to consider first the more usual situation where the species is already present at the site as the mature stage. Using this starting point, transition diagrams for all 30 different species types are presented in Fig. 4. To provide an example of how each transition diagram is derived, the diagram for an SI species can be used. (This diagram is the second in the second row of Fig. 4). An SI species type could represent a species with long lived seeds stored in the soil (S) but which is intolerant (I) and therefore must establish immediately after a disturbance.

If such a species is present at the mature stage and the site is disturbed, it will have propagules available and can regenerate and produce juveniles. The transition from the mature to juvenile stage is shown by the dashed arrow from M to J.

If no disturbance occurs to the mature stage, the individuals eventually senesce and die and, since they have the I vital attribute, they are not able to regenerate in the existing conditions. This means that the mature stage was a single age class population and, at its senescence and death, the species will appear to be lost from the site. However, propagules in the form of the long lived seeds in the soil will remain available for a period equivalent to the longevity of these seeds. Therefore in conditions of no disturbance, the mature stage of an SI species will eventually transit to the propagule stage, which is shown by the solid arrow from M to P. If still no disturbance occurs, then all the seeds will eventually die and the species will become locally extinct (as shown by the solid arrow from P to E). However, a disturbance while the species is still in the propagule stage will create the conditions necessary for an I species to germinate, establish and grow, and will lead to a juvenile stage, which, if undisturbed, will eventually develop to the mature stage. If the juvenile stage is disturb-

Table 2. The groups of species types with identical transition diagrams.

| group | species types | group | species types |
|-------|---------------|-------|---------------|
| 1 | DT, ST, VT | 9 | GR, CR, VR |
| 2 | GT, CT | 10a | ΔT, ΣT, ΓT, UT, WT |
| 3 | DI | 10b | ΔR, ΣR, ΓR, UR, WR |
| 4 | SI | 11 | ΔI |
| 5 | GI | 12 | ΣI |
| 6 | CI | 13 | ΓI |
| 7 | VI | 14 | UI, WI |
| 8 | DR, SR | | |

ed, then, due to the S character, there will still be a residual store of propagules available for regeneration, and hence the species will rapidly return to the juvenile stage. This completes the transition diagram.

The diagram for the other species types are derived in the same way. Only 14 distinctly different patterns occur in the 30 transition diagrams in Fig. 4. These are summarized in Table 2. For example, a species classified as ST has the same characteristics in a replacement sequence as one which is classified as DT. Similarly, any species with the attribute Δ, Σ, Γ, U or W in combination with T or R have the same characteristics.

The transition diagrams in Fig. 4 were derived by assuming that initially the species is established in the community at the mature stage and then examining the effects of various disturbance regimes. However it is also of interest to know what will happen if some propagules of the species reach a site which previously did not contain the species and which is well outside the normal dispersal range of the species. Transition diagrams can be drawn as above, but with the propagule stage as the starting point (Fig. 5). In this case a dashed arrow at the propagule stage indicates the outcome if the propagules arrive at a recently disturbed site while a solid arrow shows the outcome at a long undisturbed site. The diagrams are terminated at the mature stage since the transition diagrams in Fig. 4 can be derived from this point.

The transition diagrams in Fig. 5 are independent of the first vital attribute group, so only the diagrams for the T, I and R attributes are shown. It has been assumed that the arrival of propagules is a rare event and therefore

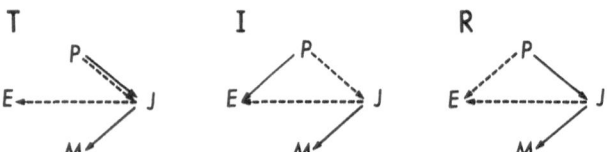

Fig. 5. Transitions between life stages at a site previously not occupied by the species. The transition diagrams start at the propagule stage (P) and terminate at the mature stage (M).

Table 3. Disturbance regimes which lead to local extinction of a species type.

| Not under any regime | Infrequent disturbances[1] | Frequent disturbances[2] | Either infrequent or frequent disturbances[3] | At first disturbance |
|---|---|---|---|---|
| 1  DT, ST, VT | 4  SI* | 2  GT, CT | 5  GI | 9  GR, CR, VR |
| 3  DI | 7  VI+ | | 6  CI | |
| 8  DR, SR | 12  ΣI* | | 13  ΓI | |
| 10a ΔT, ΣT, ΓT, UT, WT | 14  UI, WI+ | | | |
| 10b ΔR, ΣR, ΓR, UR, WR | | | | |
| 11  ΔI | | | | |

1 An interval between disturbances greater than the life span of individuals for species types marked + , and greater than life span plus propagule longevity for species types marked * leads to local extinction.

2 Second disturbance occurs before juveniles mature.

3 Interval between disturbances less than time to reach maturity or greater than life span plus propagule longevity.

the continued existence of the species at the site is dependent on its producing propagules locally.

In summary, from the 30 species types examined in Fig. 4 and Fig. 5, only 15 different patterns of transition diagrams are found. These are the 14 groups described in Table 2, but with the tenth group split into two (T and R species), since they differ in their invasive properties.

Not all of the 15 different transition diagrams are likely to be biologically feasible under particular disturbance regimes. In Table 3 the range of disturbance regimes under which various species types can persist at a site is summarized. A large group of 17 species types, representing 6 different transition diagrams can persist under any disturbance regime. In contrast, three species types, representing a single transition diagram, will become locally extinct whenever a disturbance occurs, and three other species types, each with a different transition diagram, require a particular disturbance regime to remain at a site. The remainder are vulnerable to disturbance at particular frequencies. Of these five species types will become locally extinct only if a long period elapses between disturbances, and two will become locally extinct only if frequent disturbances occur.

**Life stages**

The third vital attribute group deals with the timing of the life stages of a species. Since the timing between ⸱rtant events in the life history of an individual s determine the exact role a species will play in a ⸱cement sequence, we define three vital attributes ⸱life stage parameters) based on the life stages of a

population of a species which were described earlier. They are:

1) The time taken for a species to reach reproductive maturity after a disturbance (m).

2) The *life* span of the species in an undisturbed community (1). For T or R species the life span is effectively infinite since they form a self maintaining population. However, since I species usually form a single age class

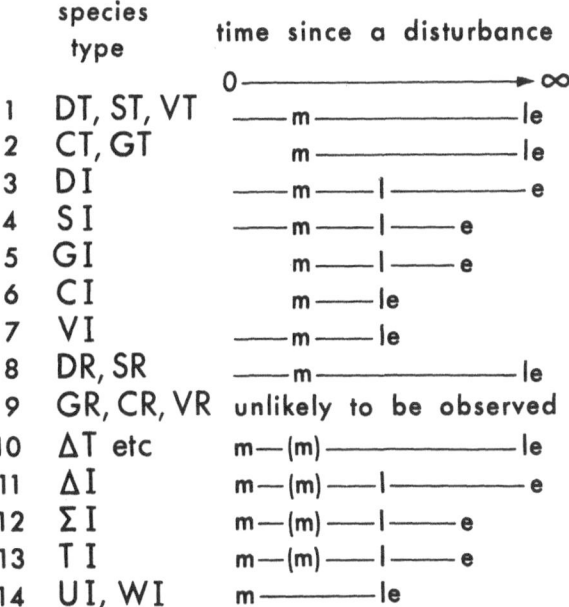

Fig. 6. Life stage parameter characteristics for each of the species types. The critical events are the time to reach reproductive maturity, m, the longevity of the species population, 1, and the longevity of its propagule pool, e. The (m) for some species types indicates the time taken to reach reproductive maturity when they invade a site previously not occupied by the species.

13

stand, the life span is approximately the longevity of the individuals within that stand.

3) The time taken for all the propagules to be lost from the community and for the species to become locally extinct (e). Local extinction cannot occur in species with indefinitely long life spans, or in species with widely dispersed seeds, but for other species the time depends on the type of propagules.

Fig. 6 shows that for most species types not all of the parameters, m, l and e, need to be known since some of the major life history events coincide. For many species types no time parameter, of only one, need be known.

### Replacement sequences

The derivation of a replacement sequence can be demonstrated by an example using the effect of fire on the wet sclerophyll and rain forests of Tasmania (Gilbert 1959, Jackson 1968). In classical terms succession progresses

from sedge and shrublands through a wet sclerophyll forest dominated by *Eucalyptus* and *Acacia* species to a mixed forest community with *Eucalyptus* overtopping a rain forest understorey. *Eucalyptus* regeneration occurs only after fire, so if no disturbance occurs the *Eucalyptus* species are ultimately lost from the community and a rainforest remains.

The two major *Eucalyptus* species, *E. regnans*, F. Muell and *E. delegatensis* R. T. Bak. are unusual in that they are among the few eucalypts which do not have the ability to form lignotubers and they rarely stem sprout. They have poorly dispersed, short-lived seed, and rely for persistence through a fire mainly on seed stored in the canopy and released from the protective seed capsules soon after a fire. They are able to regenerate only in open conditions and therefore are classified as species type CI. Another common wet sclerophyll forest species is *Acacia dealbata* Link., which can only regenerate successfully after a fire and which has seeds with long viability while stored in the soil. It is therefore of species type SI.

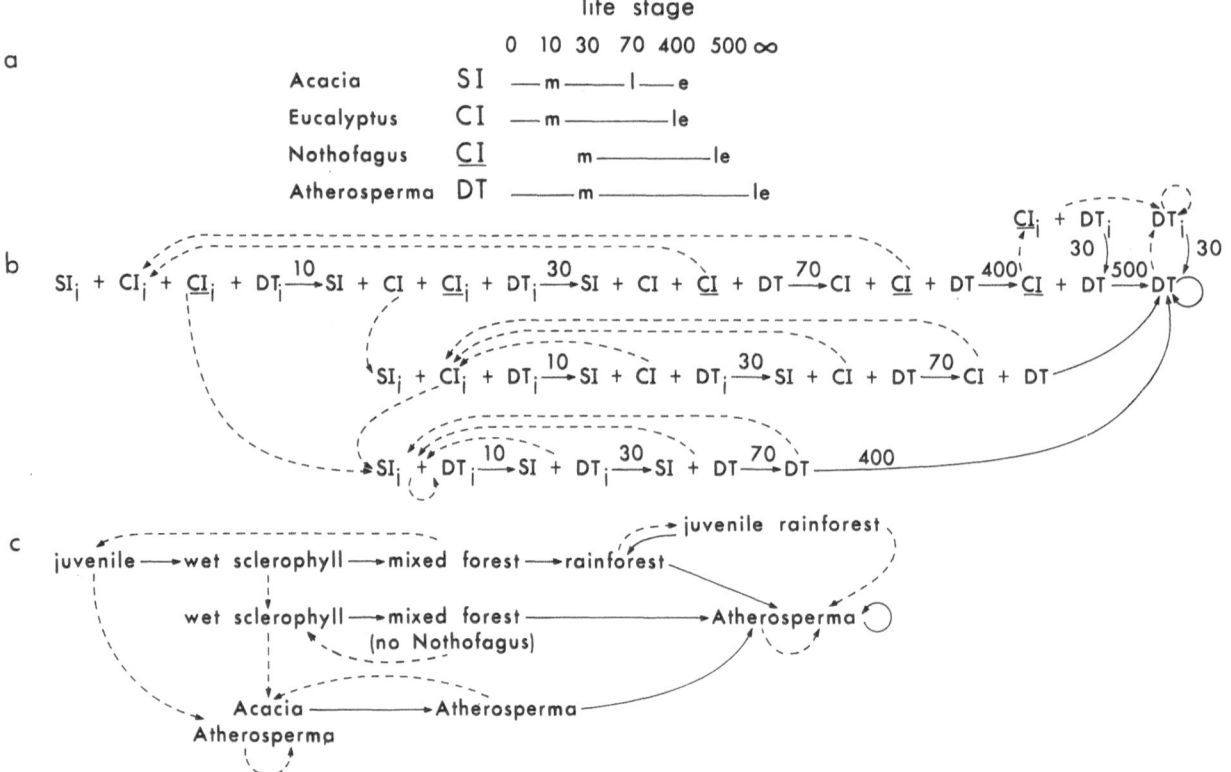

Fig. 7. The vegetation replacement sequence for the Tasmanian wet sclerophyll-rainforest system. Fig. 7a summarizes all the vital attribute data required to derive the replacement sequence shown in Fig. 7b. (The underlined CI is used only to distinguish *Eucalyptus* and *Nothofagus* in the replacement sequence.) Fig. 7c. is a summary of Fig. 7b. Solid arrows show transitions in periods with no disturbance, while broken line arrows show transitions due to a disturbance.

The two main rainforest species are *Nothofagus cunninghamii* (Hook.) Oerst. and *Atherosperma moschatum* Labill. *Nothofagus* is fire sensitive and has relatively poorly dispersed, short-lived seeds and is reliant on seeds surviving in the canopy for persistence through a fire. Regeneration of *Nothofagus* under a rain forest canopy is restricted and it is classified as a CI species. *Atherosperma* has widely dispersed seeds and it is very tolerant and is therefore a DT species.

The life stage parameters are summarized in Fig. 7a. It appears that *Nothofagus* regenerating after a fire will take about 30 years before it produces significant seed crops (Howard 1973), whereas *E. regnans* can produce seed after about 10 years (Ashton, 1956). The timing of maturity of *Atherosperma* and *Acacia* are estimated as about 30 years and 10 years respectively, although these parameters are not critical in the replacement sequence. *Acacia dealbata* has a life span of about 70 years, while its seeds may survive for another 300 to 400 years in the soil (Gilbert 1959). This means that *Acacia* seeds are available throughout the life span of the *Eucalyptus* which can live to be about 400 years old. *Nothofagus* appears to be slightly longer lived than *Eucalyptus* and has an estimated longevity of 500 years.

The replacement sequences displayed by this group of species under a variety of disturbance regimes, can be derived from the life history data in Fig. 7a. A situation where all species are present at the site in the mature life stage is taken as a starting point. If a fire occurs, then all species can regenerate via their respective propagule pools and will be in the juvenile life stage (far left of the top line of Fig. 7b). *Acacia* and *Eucalyptus* are the first to reach reproductive maturity after only about 10 years. This is shown by the first solid arrow with the superscript 10 indicating the time, measured since the most recent fire, at which the transition occurred. This is an important transition for *Eucalyptus* since it will now have a seed pool which will enable the species to persist if another fire occurs. It is less important for the *Acacia* since the species normally would have some viable seed in the soil which did not germinate or perish in the previous fire.

If no subsequent fire occurs for 30 years, the rainforest species also reach maturity. At this stage the community is essentially a wet sclerophyll forest with a *Eucalyptus* overstorey and *Acacia* and rainforest species understorey. Since *Acacia* has the I vital attribute, it will form a single age class which establishes after the fire and no further recruitment will occur. After about 70 years most individuals will have senesced and died and only

the seed pool in the soil will remain. The forest will now be a typical mixed forest with emergent *Eucalyptus* above a rainforest understorey. After about 400 years without fire, *Eucalyptus* will be lost from the community since it is also an I species like the *Acacia*. However, since *Eucalyptus* has the C vital attribute which implies a short lived seed pool, no seed pool will remain. By about this time most of the *Acacia* seeds in the soil will have become non-viable so both *Eucalyptus* and *Acacia* become locally extinct at about 400 years. This leaves a rainforest and if undisturbed over the next 100 years or so, most of the *Nothofagus* will die with very little replacement, leaving an increasingly pure *Atherosperma* rainforest.

The dashed arrows show the effect of a fire on each of the stages in the replacement sequence. If the site is burned a second time before any of the species have reached maturity (i.e. in less than 10 years), only *Acacia* and *Atherosperma* will have seed pools which will enable them to persist through this second disturbance. The *Acacia* will have some seeds remaining in the soil, while some *Atherosperma* seed will disperse from neighbouring undisturbed sites. This transition is shown by the dashed arrow from the left of the top line to the left of the bottom line of Fig. 7b. The resulting community, which includes juvenile *Acacia* and *Atherosperma* will develop as shown in the bottom line leading eventually, if undisturbed, to a pure *Atherosperma* rainforest.

The replacement sequence derived from the vital attribute data for the four species is shown in full in Fig. 7b, and summarized in Fig. 7c. If the vegetation regenerating after a fire is not burnt again then a succession from wet sclerophyll to mixed forest to rainforest and eventually to pure *Atherosperma* is derived from the vital attribute data. A fire within 10 years of a previous fire will result in the loss (or drastic reduction in stocking) of both *Eucalyptus* and *Nothofagus*, leading to an *Atherosperma* and *Acacia* forest. A fire within 10 to 30 years of a previous fire will lead to the loss of only *Nothofagus*, and a mixed forest with a pure *Atherosperma* understorey results.

Therefore this scheme produces the classical sequence of wet sclerophyll, mixed forest and rainforest, and also several other replacement sequences as well. Gilbert (1959, p. 148) commented on the mixed forests with an understorey of essentially pure *Atherosperma*, although he was uncertain about how they originated. This scheme indicates that such forests might originate from an interfire period of 10 to 30 years. Gilbert makes a similar mention of some *Acacia-Atherosperma* stands, which,

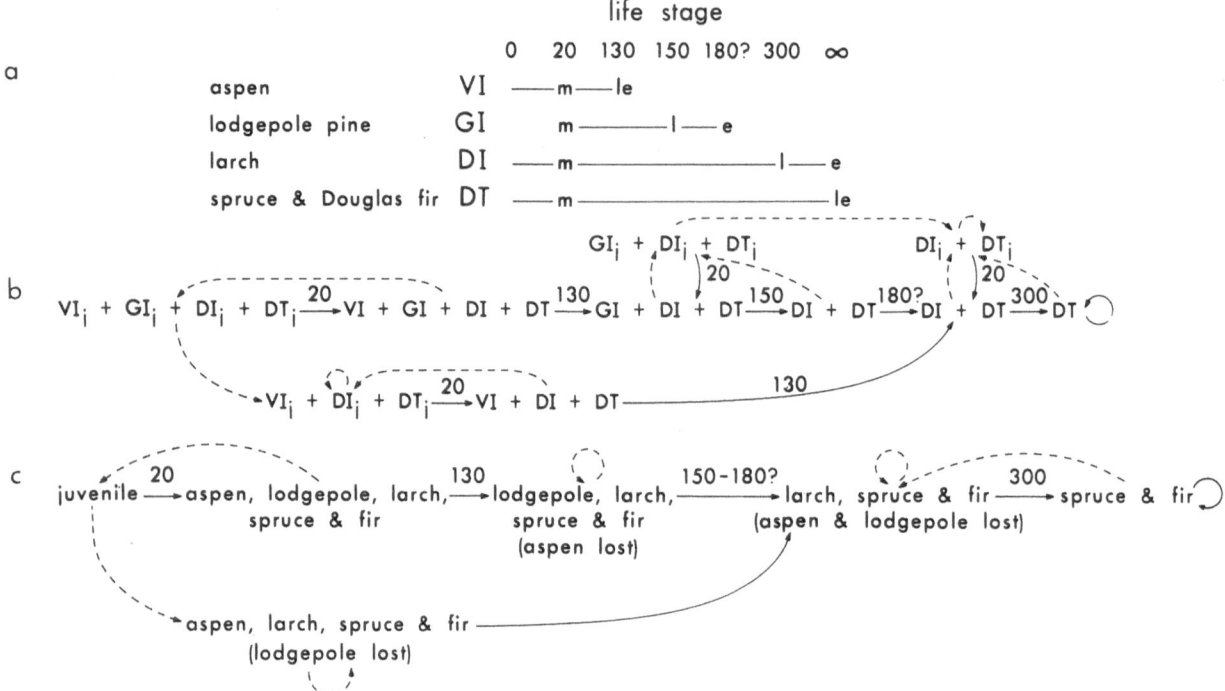

Fig. 8. The vegetation replacement sequence for an aspen community in western Montana, using the same format as Fig. 7.

according to this scheme, may arise if an inter-fire period of less than 10 years occurs.

The scheme makes possible more complete descriptions of the vegetation replacement sequence in this example than does classical succession theory. It also provides a formal framework for discussing the role of disturbance frequencies in the origin of different community types.

A second example is based on an aspen community in western Montana (Kessell 1979, Cattelino et al. 1979). *Populus tremuloides* Michx. (Aspen) will sprout after a disturbance but it is relatively intolerant and is therefore classified as a VI species. The other species include *Pinus contorta* Dougl. (Lodgepole pine) which is also intolerant, and which has serotinous cones. Seeds in these cones can survive for several decades and are shed after the cones are burnt. This is a form of long term seed storage, in which the seed pool is exhausted after the first disturbance. Therefore *P. contorta* is classified as a GI species. The other species are *Larix occidentalis* Nutt. (Larch) which is a widely dispersed, intolerant species (i.e. DI), and *Picea glauca* (Moench) A. Voss. x. *engelmannii* C. Parry (Spruce) and *Pseudotsuga menziesii* (Mirb.) Franco (Douglas fir) both of which are widely dispersed but tolerant (i.e. DT).

The full life stage attributes and the derived replacement sequences are shown in Fig. 8. The derived sequence may be summarized as follows: with inter-fire periods of 20 to 130 years a community containing a mix of all five species occurs. If the inter-fire period increases, *P. tremuloides* becomes inconspicuous in the community and soon after (at about 150 years), *P. contorta* disappears. An inter-fire period of 150 to 180 years will result in *P. contorta* regenerating from the surviving serotinous cones, but *P. tremuloides* density will be considerably reduced. If an inter-fire period of greater than 300 years occurs, *L. occidentalis* will also be lost from the community through senescence but it will regenerate when a fire eventually occurs. If very short-inter-fire periods (less than 20 years) occur, then *P. contorta* will not have time to reach maturity and restore seed stocks and it will be lost from the community, usually with *P. tremuloides* increasing in density.

These replacement sequences are in agreement with the observations of Kessell (1979) and are more comprehensive than those from a simple Clementsian successional scheme.

**Discussion**

*Variations in vital attributes*

The vital attributes of a species may vary according to

16

the circumstances in which it is found. Since the scheme deals only with the replacement sequences at a particular site we will not discuss the geographical or altitudinal variations in vital attributes shown by some species. However, it is possible that an examination of the variations in selected species along environmental gradients would prove useful, especially if the vital attributes are expressed in a more quantitative way than in this paper.

The vital attributes of species at a particular site may change under the influence of exogenous or endogenous factors. In general, exogenous factors are not included in this discussion since the scheme presented here deals with a site at which long term climatic and edaphis changes are not important in the time scale being considered. However the vital attributes of some species may change in response to the normal fluctuation in the weather experienced at a site. The most usual fluctuation will be in the second vital attribute group, and specifically from T to I and vice versa as weather conditions favouring or preventing establishment occur. Short fluctuations in weather will usually not be sufficient to change the vital attribute associated with the method of persistence or the life stage characteristics of a species. More extreme climatic events which lead to extinctions are best thought of as disturbances or catastrophes.

Several situations where endogenous factors cause a change in vital attributes can be envisaged. For example the conditions for establishment of a species may vary with the species present at a site. Consider an R species which after some delay establishes at a site, but at a density that prevents it from continuing to establish under its own canopy. In effect, it will become an I species and will die out, possibly leading to conditions where it can re-establish so showing a form of cyclic replacement.

Vital attribute changes may occur with the loss of a species which modifies the intensity of the disturbance. Again using fire as an example, if a heavy fuel producer with an I vital attribute (e.g. *Eucalyptus* species) is lost from a community then subsequent fires may be cooler and a variety of changes in vital attributes may occur. Even though we cannot know which species will show changes in vital attributes without knowing more details of the properties of those species, it is possible to predict the directions of changes that might occur. The last example above may lead to adults being able to survive fires where they could not before. The shifts in vital attributes which may then be observed are,

$$D \rightarrow \Delta$$
$$S \rightarrow \Sigma$$

$$G \rightarrow \Gamma$$
$$V \rightarrow U$$
$$C \rightarrow W$$

If the juveniles can also survive the fires, then the shifts may be

$$G \rightarrow S (\rightarrow \Sigma)$$
$$C \rightarrow V (\rightarrow U)$$
$$\Gamma \rightarrow \Sigma$$
$$W \rightarrow U$$

The bracketed transitions are included since the adults will usually also survive.

If allelopathy or forms of auto-inhibition are considered then many different changes in vital attributes can be hypothesised. Most of these changes will involve the group of vital attributes related to conditions for establishment.

However, we believe that the main emphasis should not be to find hypothetical examples of cases where vital attributes might vary because of endogenous factors, but rather to be alert for actual examples of replacement sequences which do not fit the general scheme described above. It is these examples that will provide interesting insights into the strategies of plant species in replacement sequences.

*The disturbance spectrum*

Every ecosystem is subject to a variety of types of disturbance, each with a characteristic regime. Disturbances can range in frequency from being so rare that the period is longer than the life span of the longest lived species, to being so frequent that they recur many times within the life span of the shortest lived species. A spectrum of disturbance intensities exist. At one extreme are very rare catastrophic events which alter substrates and disrupt the mechanisms of persistence and recovery. Harper (1977, p. 627) has suggested that few of the selective consequences of catastrophes are relevant to the fitness of future generations, and it is therefore unlikely that any species will have evolved mechanisms especially adapted to catastrophic disturbances. In general, highly dispersed (D) species will be the most advantaged by catastrophic disturbances, and they will usually form the pioneer species suite in the succession which follows.

In the middle of the spectrum are recurrent disturbances which may occur at irregular intervals, but nevertheless at a frequency such that there is a reasonable expectation of occurrence within the life span of many of the species within the community. Harper called these disturbances, 'disasters'; however, this seems too strong a term since recurrent disturbances are an important selective force in a population and any species which survives within a community must be adapted to these disturbances.

At the other end of the spectrum are disturbances which are effectively continuous. Factors with a continuous and significant impact, such as continuous grazing, are not strictly disturbances but are part of the normal environment, and sites which include such a disturbance should be regarded as different ecosystems from those which do not.

*Recurrent disturbances*

The scheme described in this paper is mainly concerned with recurrent disturbances. In any particular system the biota have evolved in the presence of the natural, recurrent disturbance regime. While such disturbances may be associated with local extinction of certain species or shifts in relative abundance, succession following them will generally tend to induce replacement sequences which lead to communities resembling those in adjacent undisturbed areas.

There are many different types of recurrent disturbance, including fire, flood, blowdown, extreme climatic events and disease or pest outbreaks. The vital attributes of a species are defined relative to a particular disturbance type, and therefore if a community is subject to more than one form of recurrent disturbance, more than one set of vital attributes may need to be defined for each species. However this is rarely necessary. Many disturbances do not occur independently of each other (for example, fire often follows a blowdown, and extreme climatic events weaken resistance to pests) and therefore the combination of disturbances may be thought of as the disturbance event. Also, many of the mechanisms of persistence are independent of the disturbance types (for example, D species will have propagules available regardless of the disturbance type) and therefore the vital attributes shown by the species do not vary. Similarly, the effect of many disturbance types is similar; that is, to destroy many individuals of the existing community, thereby leading to a period of reduced competition. Therefore the 'conditions for establishment' vital attribute often is unaltered.

*Continuous disturbances*

The interaction between recurrent and effectively continuous disturbances is important in understanding vegetation replacement sequences. An important example is the interaction between grazing (or herbivory in general) and fire. If grazing occurs as a series of intermittent pulses, it is best regarded as a recurrent disturbance, similar to pest outbreaks. If grazing is always present then it is part of the normal environment and it is possible to make some predictions about the way the vital attributes of some species may differ from the ungrazed situation. An extreme, but simple, example is where grazing stops the reproduction of a species. In this case all but D and $\Delta$ species will eventually become locally extinct, as the adults die and propagule pools are lost.

A more subtle example is where a species classified T in an ungrazed situation, can act as an I species in a grazed situation. This would occur if the species is sufficiently heavily grazed to prevent regeneration in normal circumstances, but after a pulse disturbance either the species regenerates in sufficient number, or the grazing animals are so reduced in number, that many individuals of the species can survive. Conversely a species can be converted from a T to an R species under the influence of grazing. This can occur if the juveniles regenerating after a pulse disturbance are heavily grazed, but, as the rest of the community recovers, some juveniles of the species survive either because other species in the community protect the grazing sensitive juveniles, or because the grazing animals feed mainly in those other areas recently subjected to a pulse disturbance. In the recovering community the grazing pressure may thus be reduced.

A species classified as an I species in an ungrazed situation can act as a T species if grazing is introduced. This often happens when grazing animals open gaps in the community allowing the species to establish in microhabitats of low competition. If the juveniles regenerating after a pulse disturbance are especially heavily grazed, then it is possible for an I species to act as an R (i.e. I→T→R). It may also be possible for an R species to act as a T species in grazed situations. This will occur if the grazing animals fulfil the requirement the species has for ecesis.

*Intensity and season of disturbance*

Throughout this paper we have not distinguished between disturbances of different intensities, but have instead used an hypothetical disturbance of an intensity 'normal' for the community. However, the intensity of many disturbances is highly variable. For example the intensity of a fire depends on the season and weather conditions, while flood, wind and other extreme weather events also vary on a continuous scale. Although it is not possible

Table 4. The manner in which vital attributes will tend to change in response to disturbances of different intensities.

| | Intensity of disturbance | | |
|---|---|---|---|
| | milder | normal | more severe |
| Method of persistence | Δ | D | D |
| | Σ | S | O* |
| | Γ | G | O |
| | W | C | O |
| | U | V | O |
| | U | U | O |
| | U | W | O |
| | Δ | Δ | D |
| | Σ | Σ | S |
| | Γ | Γ | G |
| Conditions for establishment | T | T | T |
| | O | I | I |
| | T | R | R |

* A '0' indicates that the change in intensity of disturbance leads to a loss of the mechanism responsible for the vital attribute.

to predict how a particular species will react to disturbances of greater or lesser intensity than 'normal' without some knowledge of the biology of the species, it is possible to predict the types of changes in vital attributes that might occur. These are summarized in Table 4.

The table shows the minimum change in a vital attribute that might occur, but multiple steps are possible. For example, with increasingly severe disturbances, the adults of a species with a $\Sigma$ vital attribute may lose their ability to persist through the disturbance ($\Sigma \rightarrow S$) and with more extreme disturbances the seed pool also may be destroyed ($S \rightarrow 0$). It is possible that the seed pool might be destroyed before the adults lose their ability to persist through a disturbance (i.e. $\Sigma \rightarrow W \rightarrow 0$) but this seems unlikely.

The vital attributes related to the conditions for establishment are less affected by the intensity of disturbance. It is possible that some mild disturbances might destroy the population in an I species, but not create conditions suitable for its re-establishment. In similar circumstances an R species might be able to regenerate immediately, effectively acting as a T species.

It is important to realize that some changes in the intensity of a disturbance lead to a different type of disturbance. For example, if a site is subject to flooding and silt deposition from a river every few decades, then an extreme flooding event involving fast flowing water and erosion is not a more extreme flood, but a different type of disturbance.

The season in which a disturbance occurs can affect the replacement sequence by modifying: (i) the intensity of the disturbance, (ii) the availability of propagules, or (iii) the conditions which favour the establishment of one species over another. The first effect has been dealt with above. With the second effect, only those species with D, $\Delta$ and C vital attributes are dependent on propagules produced near the time of the disturbances; S, $\Sigma$, G and $\Gamma$ species being better buffered by seed pools. Some vegetative mechanisms will be more effective in certain seasons although often they too will be independent of the season of the disturbance. Even with species with a D or $\Delta$ vital attribute, the season of disturbance is only important if it also has the I vital attribute, since T and R species are not dependent on establishing immediately after the disturbance.

The third effect is more likely to affect the relative proportions of the regenerating species rather than leading to local extinction. In this case I species will again be most affected.

In summary, it appears that the season of a disturbance is most likely to affect only the relative proportions of the species in the regenerating community. We can conceive of rare cases of local extinction due to an unusually timed disturbance (e.g. extreme weather conditions or a prescribed burn destroying a population of a CT annual before seed set), but events of this type must be so rare as to be regarded as a distinct, catastrophic, disturbance type.

*Further development*

For general ecological purposes, the main limitations of the scheme are its inability to describe which of several species, at a comparable life stage, may be dominant in terms of relative biomass or relative density. To a limited degree, this deficiency may be rectified by the addition of a fourth vital attribute group related to life form, to describe growth rate and size at maturity. The addition of such an attribute would give structure to the derived community but would enable the existing qualitative character to be retained. To account for relative density however, and for other interactions, a more quantitative description of one or more of the vital attributes may be required.

When two or more of the species at a site exhibit the same species type, or have species types which are equivalent in their role in a replacement sequence, limitations in the qualitative character of the first two vital attribute groups are evident. The interaction between three 'climax' species in the Hubbard Brook forest provides an interesting example (Forcier 1975). This scheme (Fig. 9) shows that, in the absence of disturbance beech (*Fagus grandifolia* J.F. Ehrh.) will become the major species, but it does not predict the progression from yellow birch (*Betula alleghaniensis* N.L. Brit.) to sugar maple (*Acer*

19

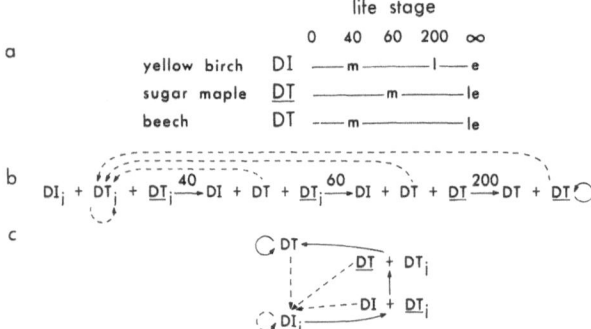

Fig. 9. The vegetation replacement sequence based on Forcier (1975). Fig. 9a and b present the vital attribute information and the derived replacement sequence, while Fig. 9c presents, in the same set of symbols, the replacement sequence described by Forcier.

*saccharum* H. Marsh.) to beech as described by Forcier. This is because, although all species have the D vital attribute, there is a ranking of dispersive abilities; yellow birch > sugar maple > beech. Therefore yellow birch tends to invade new openings and establish an initial dominance. There is also a ranking of the second vital attribute, i.e. tolerance of competition from established adults; yellow birch < sugar maple < beech. Recognition of these rankings leads to a sequence similar to that suggested by Forcier. Rather than attempt to incorporate extra degrees of detail into the present scheme, we prefer to leave it qualitative in character and general in application.

As a qualitative model, the scheme has direct application to a variety of community phenomena where presence or absence of a species is the basic information required. It has direct applicability to successional phenomena and consequently to management, particularly of natural and near-natural biomes such as rangelands and forests. It can also be applied to man-modified communities, and to such phenomena as weed invasion or suppression. In addition it may be useful as a general ecological classification system, for comparative studies of the biota of different regions, or a particular region at different times.

Applied to ecological succession, the scheme appears to have advantages over existing models based on classical views, which treat succession as a form of ontogeny. Because it is based on the vital attributes of the species present at a site, or capable of reaching it during a successional sequence, each sequence generates, in a formal manner, its own shifts in community composition. In the process, one or other notional successional pathway may be displayed, but as a result of logically determined

interactions between species types rather than as a form of community ontogeny. As a result, the scheme is flexible enough to cope with various disturbance regimes and with the removal or addition or particular species which may be associated with a particular pattern of management or disturbance.

The notions of vital attributes, species types and replacement sequences are the basic ingredients of the scheme. We hope that these ingredients will provide a framework for a variety of purposes in community biology and ecosystem management.

## Summary

A comprehensive scheme is presented which provides qualitative models of vegetation dynamics in communities subject to recurrent disturbance. The scheme has been derived to deal mainly with terrestrial communities dominated by higher plants, but may be more widely applicable.

The scheme utilizes a small number of life history attributes termed 'vital attributes' which pertain to the potentially dominant species in a particular community. Three main groups of vital attributes are recognized, relating to the method of persistence of species during a disturbance and to their subsequent arrival, to their ability to establish and grow to maturity following the disturbance, and to the time taken for them to reach critical stages in their life history.

In the application of the scheme, each major species is first categorized into a 'species type', determined by its specific attributes in the first two vital attribute groups. The interaction between various species, based on their species types and life stage attributes, then yields a replacement sequence which depicts the major shifts in composition and dominance which occur following a disturbance. Although 30 species types are recognized, only 15 distinct patterns of behaviour are displayed in replacement sequences.

Examples of replacement sequences for two different forest communities are provided.

The degree to which vital attributes are robust properties of a species is explored in relation to different disturbance frequencies and intensities, and to the seasonal time of disturbance.

## References

Ashton, D.H. 1956. Studies on the autecology of Eucalyptus regnans F.v.M. Ph. D. Thesis, University of Melbourne, unpublished.

Cattelino, P.J., I.R. Noble, R.O. Slatyer & S.R. Kessell. 1979. Predicting the multiple pathways of plant succession. Environ. Manage. 3: 41–50.

Clements, F.E. 1916. Plant succession. Carnegie Inst. Washington, Publ. 242. 512 pp.

Clements, F.E. 1936. Nature and structure of the climax. J. Ecol. 24: 252–284.

Connell, J.H. & R.O. Slatyer. 1977. Mechanisms of succession in natural communities and their role in community stability and organization. Amer. Nat. 111: 1119–1144.

Drury, W.H. & I.C.T. Nisbet. 1973. Succession. J. Arnold Arboretum 54: 331–368.

Egler, F.E. 1954. Vegetation science concepts, I. Initial floristic composition – a factor in old-field vegetation development. Vegetatio 4: 412–417.

Forcier, L.K. 1975. Reproductive strategies and the co-occurrence of climax tree species. Science 189: 808–809.

Gilbert, J.M. 1959. Forest succession in the Florentine Valley, Tasmania. Pap. Roy. Soc. Tas. 93: 129–151.

Gill, A.M. 1975. Fire and the Australian flora: A review. Aust. For. 38: 4–25.

Gleason, H.A. 1926. The individualistic concept of the plant association. Bull. Torrey Bot. Club 53, 7–26.

Golley, F.B. (ed.) 1977. Benchmark Papers in Ecology 15. Ecological Succession. Dowden, Hutchinson & Ross, Inc., Pennsylvania. 375 pp.

Harper, J.L. 1977. Population biology of plants. Academic Press, London. 892 pp.

Horn, H.S. 1976. Succession. In: R.M. May (ed.), Theoretical Ecology: Principles and Applications. p. 187–204. Blackwell, Oxford.

Howard, T.M. 1973. Studies on the ecology of Nothofagus cunninghamii Oerst., 1. Natural regeneration on the Mt. Donna Buang massif, Victoria. Aust. J. Bot. 21: 67–78.

Jackson, W.D. 1968. Fire, air, water and earth – an elemental ecology of Tasmania. Proc. Ecol. Soc. Aust. 3: 9–16.

Jarrett, P.H. & A.H.K. Petrie. 1929. The vegetation of Black's Spur region, II. Pyric succession. J. Ecol. 17: 249–281.

Kessell, S.R. 1979. Gradient Modeling. Springer-Verlag, New York. 320 pp.

Lyon, L.J. & P.F. Stickney. 1977. Early vegetational succession following large northern Rocky Mountain wildfires. Proc. Tall Timbers Fire Ecol. Conf. 14: 355–375.

Naveh, Z. 1975. The evolutionary significance of fire in the Mediterranean region. Vegetatio 29: 199–208.

Noble, I.R. & R.O. Slatyer. 1977. Post fire succession of plants in Mediterranean ecosystems. In: H.A. Mooney & C.E. Conrad (eds.), Proc. Symp. Environmental Consequences of Fire and Fuel Management in Mediterranean Ecosystems, pp. 27–36. U.S.D.A. Forest Service Gen. Tech. Rep. WO-3.

Noble, I.R. & R.O. Slatyer. 1978. The effect of disturbances on plant succession. Proc. Ecol. Soc. Aust. 10: 135–145.

Accepted 31 October 1979

## Appendix

*Definitions*

| | |
|---|---|
| Propagule: | A structure produced by an organism, either sexually or asexually, which becomes detached from the parent and gives rise to another individual. |
| Vital attribute: | An attribute of a species which is vital in determining its role in vegetation replacement sequences. |
| Vital attribute group: | A set of related vital attributes. |
| Mechanism: | A biological mechanism that is wholly or partly responsible for a vital attribute displayed by the species. |
| Life stage: | The juvenile, mature, propagule or extinct stage of development of a species population. |
| Pattern of availability: | Those life stages of a species during which a vital attribute is effective. |
| Species type: | A description of a species in terms of the first two vital attribute groups. |
| Replacement sequence: | The gain or loss of species (not necessarily on a one to one replacement basis) from a community with the passing of time. |
| Species presence: | The presence of sufficient individuals (or propagules) to provide a continuing, viable gene pool, providing the environmental conditions remain suitable. |

# VEGETATION DYNAMICS AND SEX STRUCTURE OF THE POPULATIONS OF PIONEER DIOECIOUS WOODY PLANTS

Janusz B. FALIŃSKI

Geobotanical Station, 17.230 Białowieza, Poland

**Keywords:**

Dioecious woody species: *Juniperus communis*, Old farmland, Old meadow, Population dynamics, Population structure *Populus tremula*, *Salix*, Sex ratio, Vegetation succesion

## Introduction

The incursion of trees and shrubs in the course of succession is frequently so impetuous that it has often been referred to as 'invasion' (e.g. Blackburn & Tueller 1970, Bråkenhielm 1977, Tüxen 1973, Hard 1975, Meisel 1978, Faliński 1980, in press).

Studies on vegetation succession lead to the question – what the causes are of the exceptional success attained by certain tree and shrub species in the establishment on new territories (Faliński 1980).

Data from the literature and the author's own observations indicate that the list of species invading old fields and meadows and city ruins as well as recently denuded areas, for instance in the process of deglaciation, includes representatives of the same few genera: *Salix*, *Populus*, *Betula*, *Alnus*, *Pinus* and *Juniperus*. What are the common features uniting these species?

With the exception of *Juniperus* all these species are anemochorous. Further comparative analysis revealed further biological and ecological feautures of the species of the above-mentioned genera (Fig. 1). These are: a strong tendency to vegetative propagation, together with effective generative multiplication, anemogamy, that is wind-pollination, with the exception of *Salix*, dioecism, with the exception of *Alnus Betula* and *Pinus* which are monoecious and monosexual, and proanthia, that is flowering before development of foliage, with the exception of *Pinus* and *Juniperus*. To the list of these characteristics should be added: a tendency to mass occurrence and agglomeration, low nutritional requirements, and a short individual life span with relatively early maturation for generative reproduction. The five main characteristics first enumerated are mostly fully represented in *Populus* (Fig. 1). In the pertinent literature attempts may be found to analyse the role of some of the factors mentioned, for instance anemochory. From the ecological point of view the role of dioecism, which is rather common among pioneer species, is the most obscure (Faliński 1980).

Dioecism (in plants not as common as in animals) allows to consider a sex structure of some populations, for instance in the pioneer woody species. The sex structure of the dioecious species populations is characterized by two indices:

1. Participation of reproducing individuals in a population in comparison to all individuals, or to not-reproducing individuals (juveniles, seniles).

2. Participation of male or female individuals in the total number of reproducing individuals, i.e. the ratio of male individuals to female ones, or sex ratio; it is usually shown as a formula $\male : \female = 1 : \ldots$.

In stable populations the sex ratio is mostly $1:1$. However, deviations are well known (Freeman, Klikoff & Harper 1976, Harper 1978, Vernet 1971).

| | Anemochory | Strong tendency to vegetative propagation | Anemogamy | Dioecism | Proanthy | Σ |
|---|---|---|---|---|---|---|
| POPULUS | ● ● ● | ● ● | ● ● ● | ● ● ● | ● ● ● | 5 |
| SALIX | ● ● ● | ● ● ● | entomogamy | ● ● ● | ● ● | 4 |
| JUNIPERUS | ornithochory barochory | ● ● | ● ● ● | ● ● ● | evergreen | 3 |
| BETULA | ● ● | ● | ● ● ● | monoecious | ● ● | 4 |
| ALNUS | ● ● | ● ● | ● ● ● | and | ● ● | 4 |
| PINUS | ● | ● | ● ● ● | monosexual | evergreen | 4 |

Fig. 1. Biological characteristics of the woody pioneer plants (after Faliński, 1980).

The search for relationships between the sex structure of the population and vegetation dynamics or vegetation differentiation was rarely attempted and then only for herbaceous plants (Vernet, 1971, Zarzycki & Rychlewski 1972, Falińska 1979).

Investigations on the proportion of male and female individuals in populations of dioecious woody species are rather numerous. They dealt, for instance, with *Populus tremula L.* (Langhammer quoted by Białobok 1973), *Populus tremuloides* Michx (Muhle Larsen, 1970, Pauley & Mennel 1957, Framer 1964), *Salix viminalis* (Rabotnov, 1958) and *Acer negundo* (Freeman, Klikoff & Harper 1976).

Similarly, as in herbaceous plant populations, the variability of the sex ratio in populations of woody plants was considered mainly to be the result of a simple habitat or for instance of competition (Zarzycki 1975), and genetic conditioning. The reviews of Freeman, Klikoff & Harper (1976) and Zarzycki (not published) suggest that these factors operate on a *syn*dynamic basis. They might, perhaps at least partly, be interpreted as the effect of

vegetation succession or at least connected with succession. This does not rule out, that some of the investigators, are of the opinion that there is an influence of habitat conditions on the sex ratio.

The present paper is an attempt to present the changes in sex structure of pioneer tree and shrub populations with reference to regeneration succession on old farmland and old meadows in the Białowieża Primeval Forest region (Faliński 1966, 1972, 1977).

This process is an exceptionally rewarding object for research, because in one succession series, e.g. an oligotrophic pine forest habitat – *Peucedano-Pinetum typicum* – as many as 5 dioecious woody species are found (*Juniperus communis L., Populus tremula L., Salix rosmarinifolia L., S. aurita L., S. caprea L.*); sporadically other willow species are found. Monoecious species (*Pinus silvestris, Betula pendula*) appear in later stages. Similarly, in the early stages of vegetation development on old fields, meadows, clearings and forest edges in *Alnus glutinosa* bog forest, and mixed forests, *Salix pentandra, S. cinerea, S. aurita,*

Fig. 2. The main study object: old farmland at oligotrophic pine forest habitat, colonised by juniper. 6 succession phase. (photo J. Hereźniak).

24

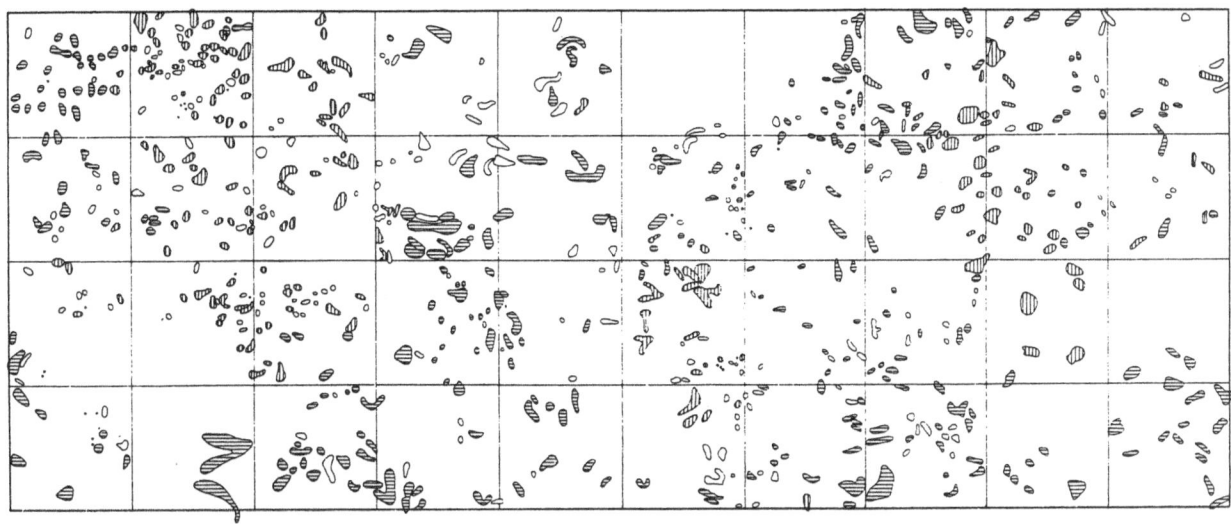

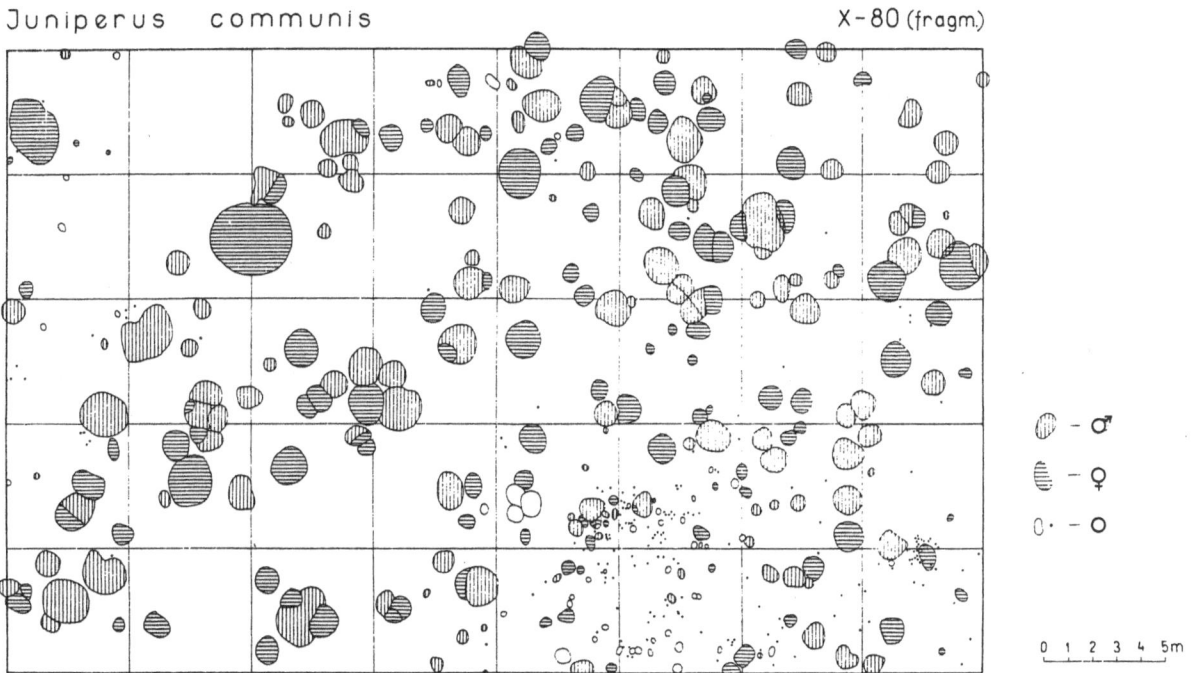

Fig. 3. Example maps of spatial sex structure of pioneer dioecious scrub species populations. Each permanent plot measures 1000 m².

*S. rosmarinifolia, S. starkeana, S. myrsinifolia, S. caprea* and their numerous hybrids appear. (Figs 2, 3 and 4).

**Materials and methods**

The basic investigations were performed in habitats on poor soil, primarily overgrown with pine forest of the *Peucedano-Pinetum typicum* and additionally on: *Cladonio-Pinetum, Carici elongatae-Alnetum, Circaeo-Alnetum, Tilio-Carpinetum* and *Pino-Quercetum* habitats (Figs 2, 3, 4). Studies on permanent plots made it possible to distinguish the main phases of vegetation development and unite

Fig. 4. Sex structure of the *Salix cinerea* population in the valley of the river Leśna. (photo J.B. Faliński).

them into one succession series. The duration of the particular phases and *absolute* dating were based on the knowledge of the age of *Juniperus communis* and other tree and shrub populations. In other habitats dating was based on determinations of the age of *Salix*. The data were verified by interviews with the local people, forest service and analysis of cadastre maps. The scheme of vegetation succession on oligotrophic pine forest habitats is presented in Fig. 5.

Analysis of changes in the population structure of pioneer tree and shrub species in connection with vegetation succession was based on samples of populations of these species collected on the above mentioned permanent plots. The basic experimental plot was 1000 m² in size. Analyses are based on sex and age pyramids and pyramids of sex and size (Figs 6–10). The age of the individuals was dated by dendro-chronological methods, and by counting the year rings. The population age was determined by the age of the oldest individuals in it. The age of the

studied populations was within the range of 5 to 55 yr. Changes in the size structure, sex structure and changes in population density, as well as vegetation succession, have been followed simultaneously for several years on twin permanent plots (Table 1).

**Results**

1. Regeneration vegetation succession in pine forest, *Peucedano-Pinetum-typicum*, habitats from fresh fallow land to the phase of older *Juniperus Populus tremula* brushwood takes place over about 70 yr. Its course is anthropogenically retarded by extensive sheep grazing (in early phases) and plunder felling, mainly of *Pinus* (in the final phases). beginning of propagation phase,

2. Synchronisation of vegetation development (vegetation succession) and development of the main species involved in this process, i.e. *Juniperus*, are shown in Fig. 14.

26

| 0 | 1 | 2 | 3 | 4 | 5 | 6 | 7 | 8 | 9 |
|---|---|---|---|---|---|---|---|---|---|
| Segetal community Teesdaleo-Arnoseridetum | Pioneer phase of Koelerion glaucae with segetal species | Perennial grasses, herbs and lichens communities: Koelerion glaucae, Cornicularia - Cladonietum, etc. | | | Complexy, herbs communities: Koelerion, Cornic.- Cladonietum, Sperg. Corynephoretum. Juniperus brush comm. | | Juniperus –Populus brushwood | communis- tremula community | Spontaneous pine forest Peucedano- -Pinetum |
| | | free of juniper | with pioneer junipers | with sex de- term. junipers | earlier phase | optimal phase | earlier phase | later phase | earlier phase |
| | 1-2 | 2-5 | 5-12 | 8-15 | 12-18 | 15-25 | 25-45 | 40-70 | > 70 |

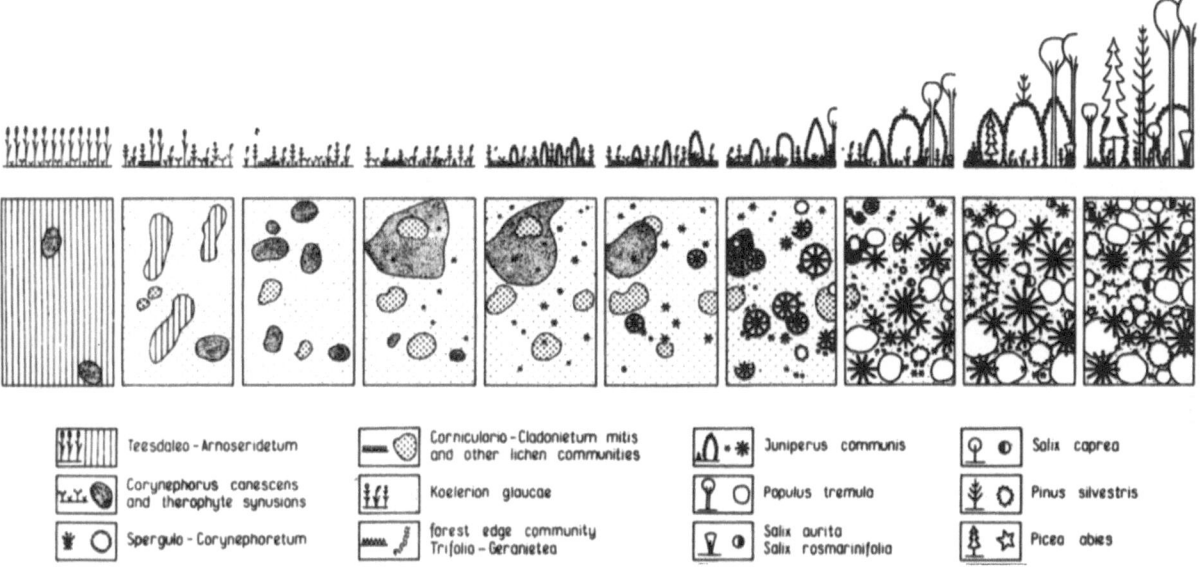

Fig. 5. Scheme of the course of plant succession in old farmland.

The succession process is described in 8 development phases which may be combined into 3 stages. The initial stage comprises: therophyte communities, lichen and sandy grassland communities; the optimal stage includes: a complex of grassland and *Juniperus* brush; the terminal stage brings juniper-aspen brushwood. To the particular developmental stages correspond the following phases of the *Juniperus* population:

– *initial succession stage*: invasion and colonisation phases,
– *optimal succession stage*: – stabilisation phase and beginning of propagation phase,
– *terminal stage* – phase of intensive propagation, over-condensation and regression of population. This stage is at the same time the initial stage for *Pinus* forest development (Figs. 5 and 14).

3. The main phases of *Juniperus communis* population development can be followed in the particular sex and age pyramids of the oldest populations (Fig. 6) and by comparison of pyramids for populations of various ages which arose and developed under analogous habitat conditions (Fig. 7).

In the first case, beginning from the vertex, traces of the following phases can be reconstructed (Fig. 6):

age 55–45 yr (1922–1931) colonization
45–30 yr (1932–1947) stabilization
29–12 yr (1948–1964) stabilization and propagation
12– 3 yr (1965–1974) overcondensation and regression.

4. Comparison of *Juniperus communis* populations participating in the successive phases of vegetation succession on fields, as well as direct observations for several years on permanent plots (Table 1), reveals changes in the sex structure of these populations. These changes are characterised not only by a differing contribution by generatively reproducing individuals, but also by changes in sex ratio (Fig. 7).

Younger *Juniperus communis* populations in old fields with a small proportion of reproducing individuals

Table 1. Changes of density and sex structure in the juniper population on the oldfield (examples).

| Phase of vegetation and population development | Number of permanent plot | Sex, sex indices | Individuals/ 0.1 ha in year | | | | |
|---|---|---|---|---|---|---|---|
| | | | 1975 | 1976 | 1977 | 1978 | 1979 |
| 4 Perennial grasses, herbs and lichens communities – with sex determ. juniperus | 66.13 | ♂ | 8 | 9 | 33 | 45 | 62 |
| | | ♀ | 7 | 10 | 20 | 28 | 36 |
| | | O | 181 | 214 | 193 | 181 | 175 |
| | | Σ | 196 | 233 | 246 | 254 | 273 |
| | | $\frac{♂ + ♀}{Σ} \cdot 100\%$ | 7.7 | 8.2 | 21.5 | 28.7 | 35.9 |
| | | ♂ : ♀ | * | * | 1:0.6 | 1:0.6 | 1:0.6 |
| 5 Complexy of herbs and lichens communities with juniperus brush community – earlier phase | 66.17 | ♂ | ** | 18 | 29 | 33 | 35 |
| | | ♀ | | 15 | 24 | 29 | 32 |
| | | O | | 44 | 31 | 25 | 28 |
| | | Σ | | 77 | 84 | 87 | 95 |
| | | $\frac{♂ + ♀}{Σ} \cdot 100\%$ | | 42.9 | 63.1 | 71.3 | 70.1 |
| | | ♂ : ♀ | | 1:0.8 | 1:0.8 | 1:0.9 | 1:0.9 |
| 6 Complexy of herbs and lichens communities with juniperus brush community – optimal phase | 66.1 | ♂ | 83 | 80 | 88 | 96 | 109 |
| | | ♀ | 80 | 85 | 88 | 97 | 105 |
| | | O | 77 | 91 | 104 | 91 | 94 |
| | | Σ | 240 | 256 | 280 | 284 | 308 |
| | | $\frac{♂ + ♀}{Σ} \cdot 100\%$ | 67.9 | 64.4 | 62.9 | 67.9 | 69.4 |
| | | ♂ : ♀ | 1:1.0 | 1:1.1 | 1:1.0 | 1:1.0 | 1:1.0 |

* calculation non rational ** no data
O 'neutral' = non flowering individuals (juvenile or senile)

exhibit a certain preponderance of male individuals. This is the consequence, among other things, of the earlier maturity for reproduction of male- as compared with female individuals. The above – mentioned preponderance is no longer noted in stabilized populations ♂ : ♀ = 1 : 1), and in older populations a gradual prevalence of female individuals is observed. Thus reliable sex ratio changes lie within the limits of ♂ : ♀ = 1 : 0,4 to 1 : 1,15.

5. The relatively late appearance in the succession process of the *Populus* and *Salix* species (phase 6) allows analysis of their population sex structure only in the terminal stage of vegetation development. This corresponds to the phase of overcondensation and recession of the *Juniperus* vegetation.

Populations of all three *Salix* species (Fig. 8) present at that time consist exclusively of individuals reproducing generatively. The rare juvenile individuals are mostly of vegetative origin. The youngest flowering individuals were 5 years (*S. rosmarinifolia*) and 7 years (*S. aurita*) old. In both cases they were female (Fig. 8).

The samples sizes allow only approximate determinations of the sex ratio:

for *Salix rosmarinifolia* ♂ : ♀ = 1 : 3.2 at population age 21 yr

for *Salix aurita* ♂ : ♀ = 1 : 4.3 at population age 32 yr

for *Salix caprea* ♂ : ♀ = 1 : 2.0 at population age ca 20 yr.

Sex manifested in *Populus tremula*, in contrast to *Juniperus* and *Salix*, by a sudden flowering of a large part of the population at the same time (Fig. 9). In comparable samples from 1000 m² plots with about 400

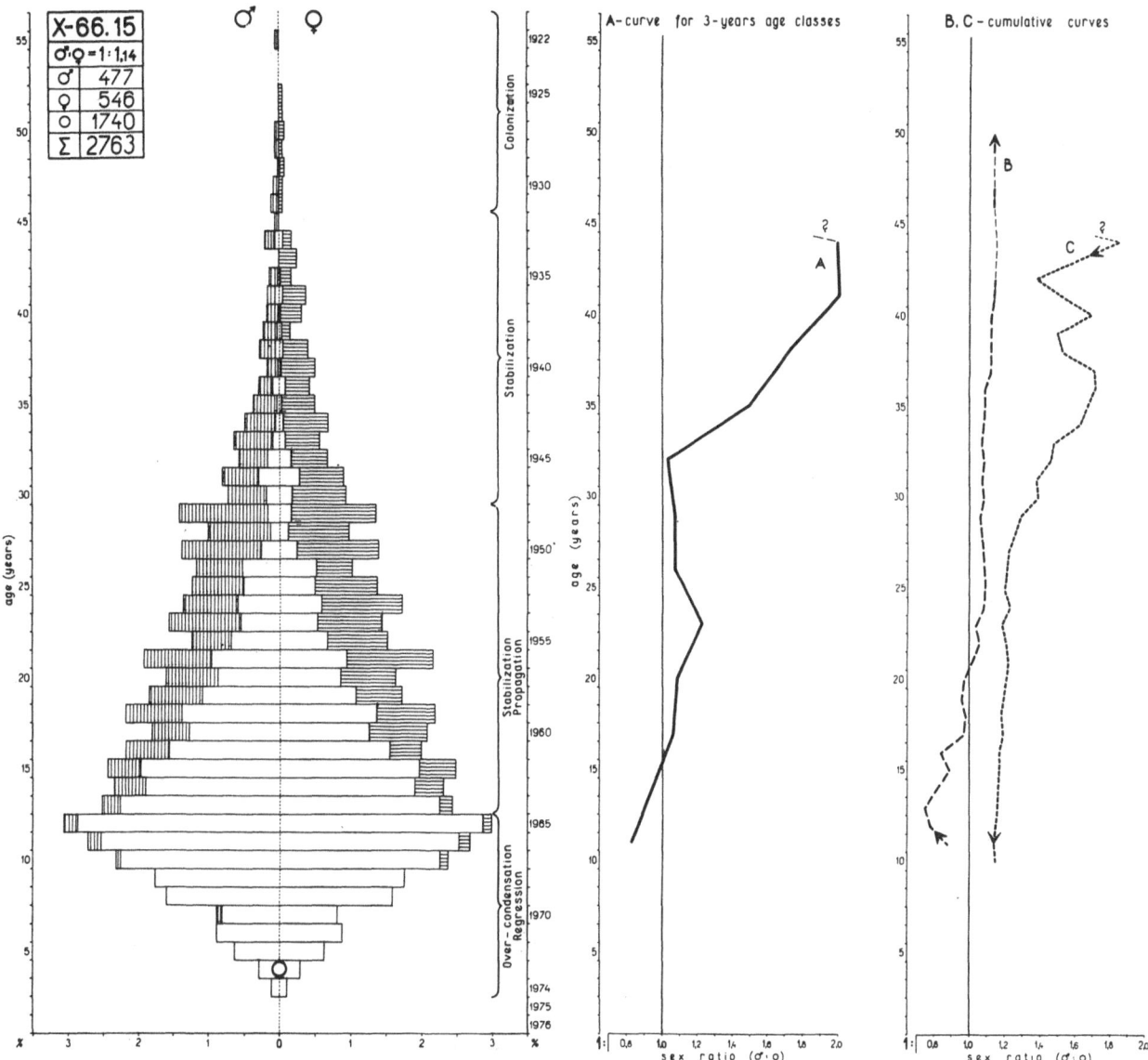

Fig. 6. Sex and age pyramid in the oldest population of *Juniperus communis* (55 yr old) in the phase of increasing propagation and over-crowding, i.e. in terms of plant succession in the older phase of juniper-aspen communities. Within the range of the pyramid: reconstruction of main phases of the development of the population. Next: an analysis of the sex structure of the pyramid (the margin edge) classes are omitted due to low numbers;

A- changes of the sex ratio in consecutive 3 yr age classes;

B- cumulative curve of the sex ratio coefficient with consecutive (successive) addition of the increasingly reproducing individuals from the bottom to the top of the pyramid, e.g. 10 years old, 10 + 11, 10 + 11 + 12 etc./;

C- cumulative curve of the sex ratio coefficient with successive addition of increasingly reproductive individuals from the top of the pyramid to the bottom (e.g. 44 yr old, 44 + 43, 44 + 43 + 42, etc.).

individuals a marked prevalence of male individuals appears immediately. The sex ratio reaches the value ♂ : ♀ = 1 : 0.01, thus one female individual per 100 male ones or more. This ratio in pioneer populations of *P. tremula* may change only slightly with time, where female individuals start to reproduce.

29

The *P. tremula* populations examined so far in 11 samples can be decidedly classified as 'male'. Attainment of the theoretical ♂ : ♀ = 1 : 1 ratio in older populations, found for instance in the forest populations of Norway, is only possible after a large proportion of male individuals die out. It cannot be ruled out, however, that in some regions on old farmland 'female' populations may still be found (Fig. 10).

Thus, over a larger area we would be dealing with a 'patchy' population space structure with regards to sex ratio index.

6. The changes in sex structure, particularly those in sex ratio connected with succession, observed in the succession series on pine forest habitats in *Juniperus communis* and *Populus tremula* populations were also observed in the populations of several *Salix* species involved in vegetation regeneration in other habitats (Fig. 11).

It was obvious that both increase of the density and changes of the plant community structure in the final vegetation succession stage on an old farmland, are the main factors responsible for changes of the sex structure and the size structure in older populations of *Juniperus communis* (Figs. 12 and 13; for a more detailed description see Discussion).

## Discussion

At present the action of at least 3 factors responsible for the changes in the sex structure of the pioneer populations of woody species in the course of succession can be considered. In the order of their occurrence in the course of succession they are: (1) The state of the habitat in old farmland in the initial stage of succession, or rather: in

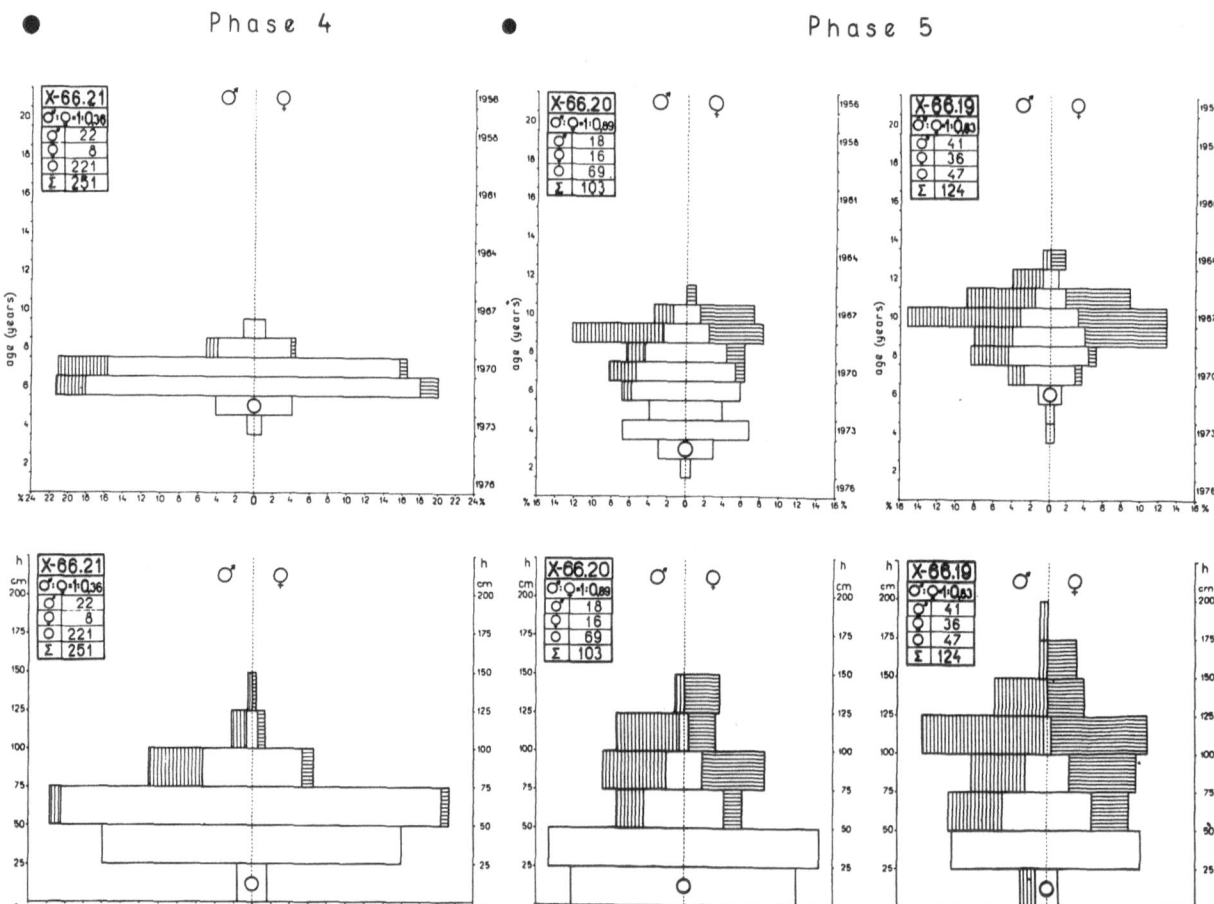

Fig. 7. Pyramids of sex and age and sex and size sampled taken from the population of *Juniperus communis* at the phase of colonization (phase 4), the younger phase of stabilization (phase 5) and the older phase of stabilization (phase 6).

the phases of invasion and colonization of the habitat by *Juniperus*; (2) differentation of vegetation and habitat in the optimal stage of succession; and (3) overcondensation of the *Juniperus* population and of the plant community (covering) in the terminal succession stage (phases 7 and 8).

The pioneer conditions are determined by the naturally poor habitat, which is secondarily still further impoverished, and to some extent transformed by extensive agricultural management for many centuries. These conditions do not only retard the spontaneous development of vegetation and colonisation of old fields by *Juniperus*, but may also act selectively at the moment of germination of diaspores of this shrub. The mechanism and causes of such selection, are not yet known, although the characteristics of old farm soils are by now well known (Strzelecki & Sobczak 1972). We can, therefore, only presume

that this selection would reach the diaspores and seedlings to a greater extent from which female individuals will develop.

This leads to the further assumption that the male individuals in pioneer species are more resistant to extreme conditions already during the germination and in the juvenile phase. Such is the situation, for instance, for the shrub *Salix cinerea*. In the Białowieża Primeval Forest, under conditions of periodical flooding on low peatfields in the river valley, the male clones occupy the deeper troughs (unpublished observations). Thus they tolerate a longer water stagnation than female individuals do. The same counts for *Salix pentandra*.

The transformation of the habitats in the course of succession leads, among other things, to a shallowing or even complete levelling of small hollows, which creates conditions favourable for the appearance of female

Phase 6

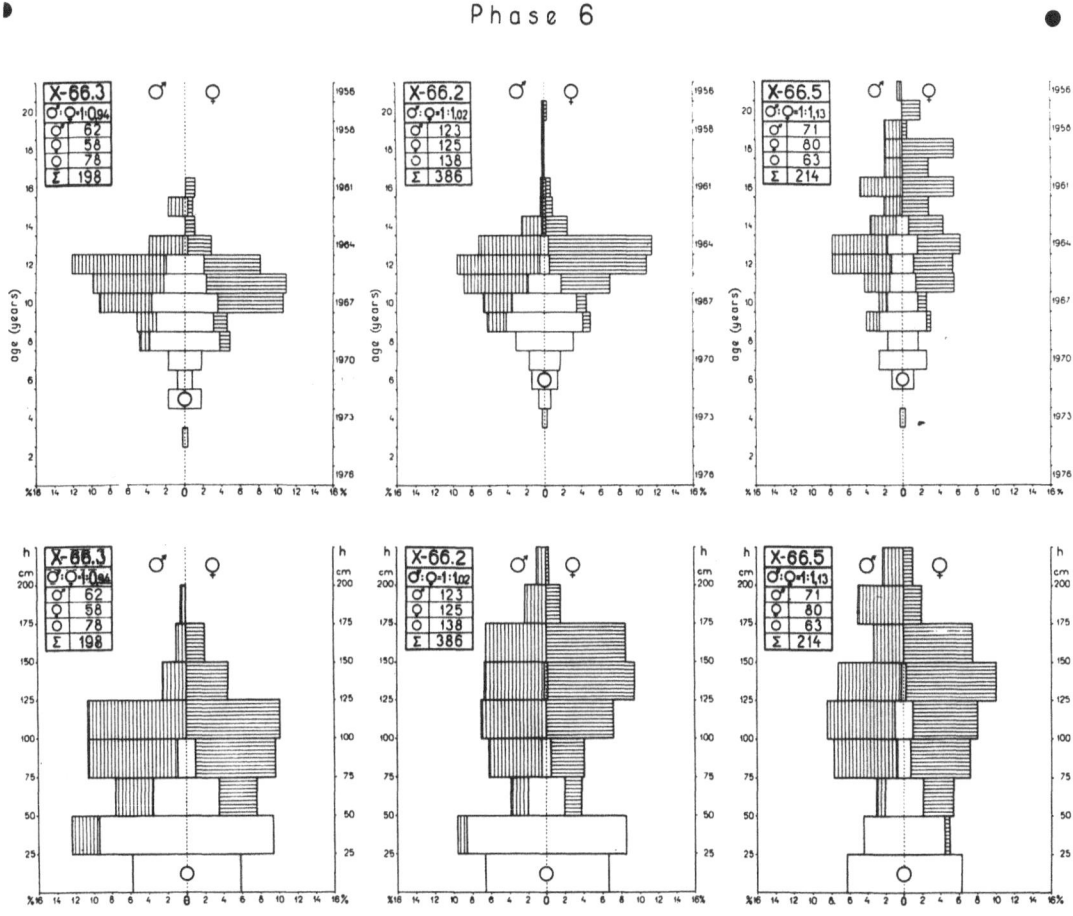

31

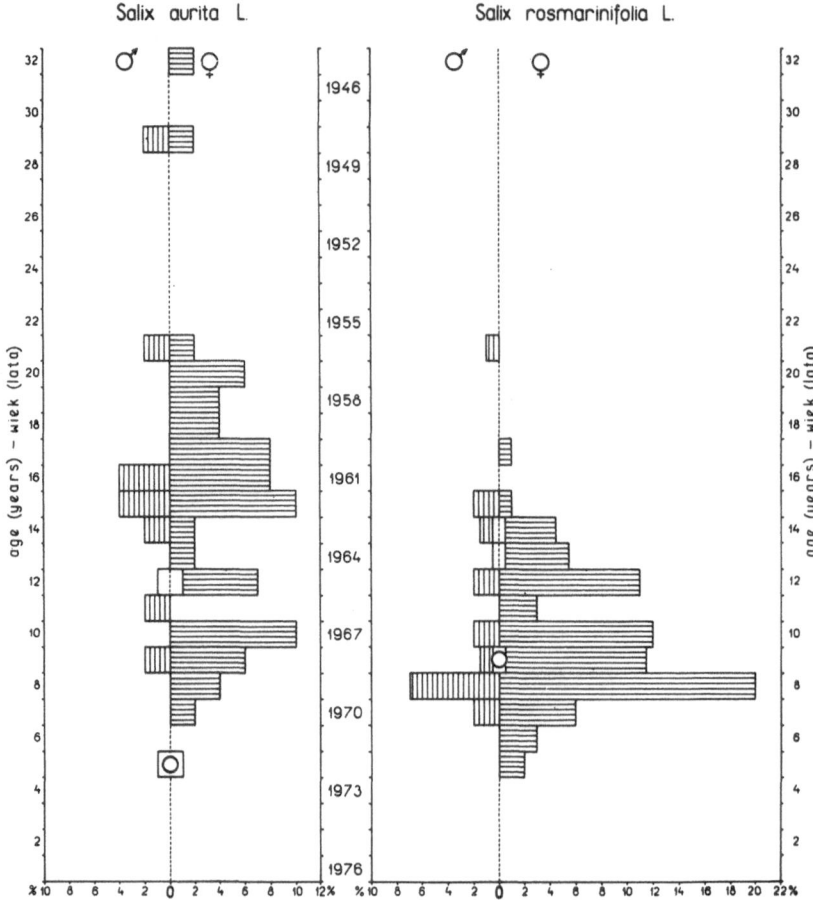

Fig. 8. Examples of the sex and age pyramids from populations of *Salix aurita* and *S. rosmarinifolia*.

individuals. A certain analogy may be also found here in the occurrence of male and female individuals of the large tree *Salix caprea*. This species appears as a high-growing tree particularly in the regenerative phases of mixed forest of the *Tilio-Carpinetum* type, e.g. along roads through the forest. In the neighbourhood of the changed or denaturated stands this willow species is relatively more frequently represented by male individuals, although exceptionally the contribution of male and female sexes may be equal (unpublished observations).

Our findings concerning the agglomeration of a greater number of male individuals when conditions tend to be extreme, for instance conditions of drought or high salinity, which are specific mainly, though not exclusively, for pioneer habitats, agree with the reports of Freeman, Klikoff & Harper (1976) and also of Putwaih & Harper (1972). I am of the same opinion as Putwain & Harper

(1972) and Zarzycki (unpublished) that in these cases one may even speak of separate ecological niches of male and female individuals.

The gradual decrease of extreme conditions in the course of the succession, caused for instance by accumulation of organic-matter in the soil, progress of soil processes and differentiation of the structure of plant communities does not only mean a change in the conditions under which the population lives. It may also favour changes in the numerical proportions of male and female individuals in the populations, that is· the sex ratio.

Let us now consider in detail the changes in sex structure of *Juniperus* populations in connection with the drastic increase in density, characteristic for the end phases of succession. Special evidence of the influence of density on the sex structure and population size structure of *Juniperus communis* is supplied by analysis of analogous age classes

Populus tremula L.

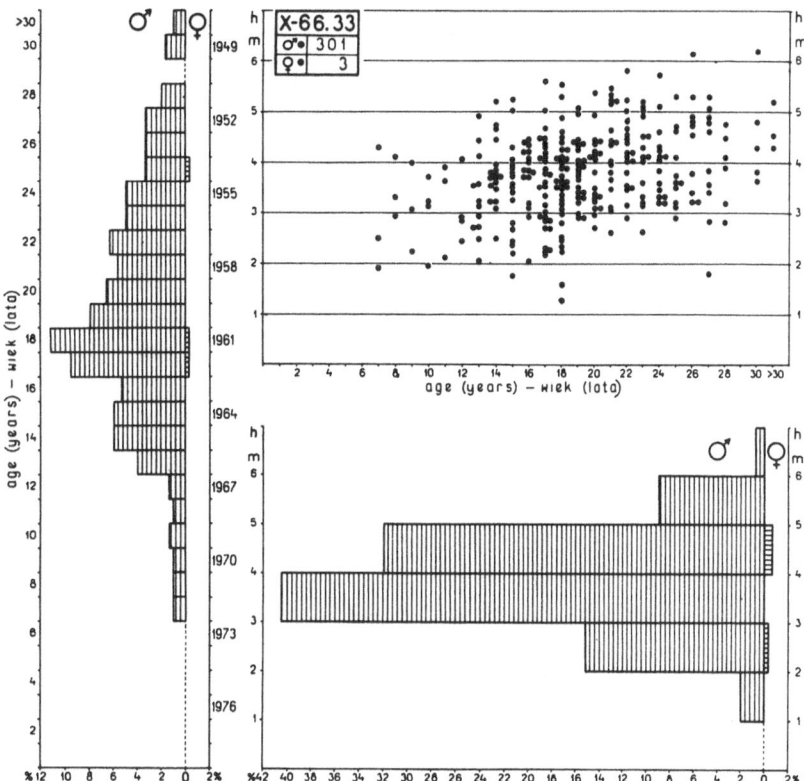

Fig. 9. Pyramids of sex-and-age and sex-and-size for chosen samples from the population *Populus tremula*. Only those individuals which flowered during 1977 and 1978 in the area of 1000 sq. m. were chosen.

Populus tremula L.

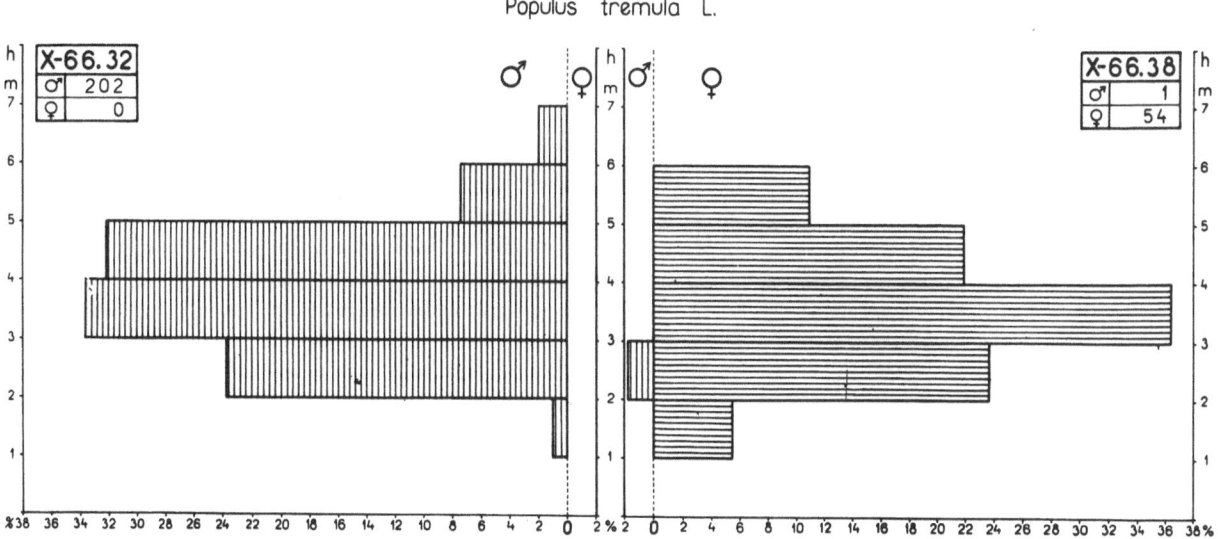

Fig. 10. Sex and size pyramids of 'male' and 'female' populations of *Populus tremula*. Only those individuals which were flowering during 1977 or 1978 were taken into account. Each sample was taken from an area of 1000 sq. m.

33

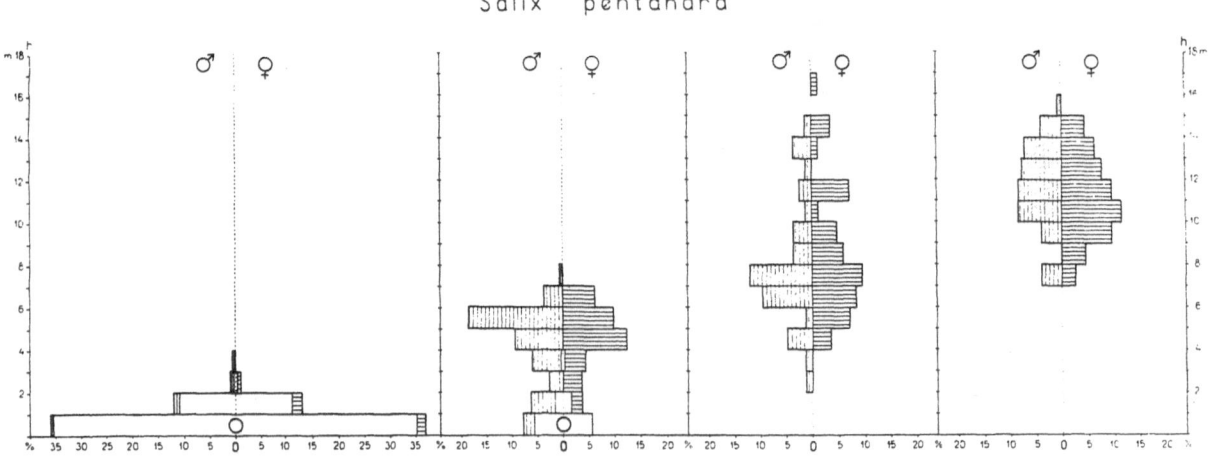

Salix pentandra

Salix cinerea

| Suposed age of population | ≈5 years | ≈15 years | ≈20 years | ≈30 years |

Fig. 11. Pyramids of sex and size from populations of *Salix pentandra* and *Salix cinerea*.

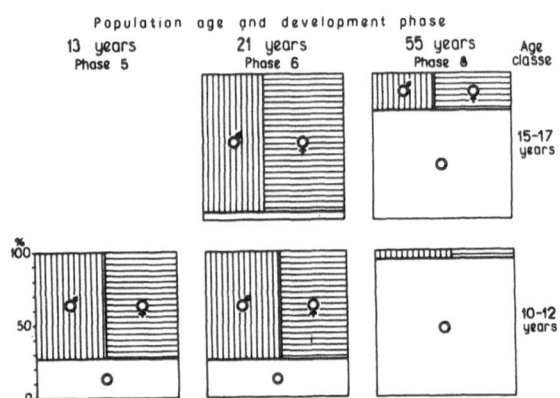

Fig. 12. Comparison of the sex structure of analogous age classes (10–12 yr) and (15–17 yr) of the younger (13 yr old/older and the oldest/55 yr old) populations of *Juniperus communis*, corresponding to the successive phases of the development of the population and phases of the plant succession. Data are taken from the surface samples no.: 66.12, 19, 21, 2, 3, 5, 15.

in different-aged populations, occurring in various phases of vegetation succession (Figs. 12 and 13). Age classes of 10–12 and 15–17 yr were subjected to analysis in populations aged 13, 21 and 55 yr. Individuals in the populations 13 and 21 yr old in the phase of stabilisation (phases 5 and 6) with a density of up to 3900 ind./ha. are dispersed over the stand. In the oldest population with a density of *Juniperus* alone up to 30 000 ind./ha. that is almost 10 times higher, other tree and shrub species are present: *Populus tremula*, *Pinus silvestris*, *Betula pendula*, *Salix caprea*, *S. aurita*, *S. rosmarinifolia*, *Cytisus ruthenicus*. This situation corresponds to the older phase of juniper-aspen brushwood.

In the 10–12 yr age class, both in the 13 – and 21 – yr population, more than 70 % of the individuals participate in generative reproduction (Fig. 13). On the other hand, in the oldest population the contribution of reproductive individuals in the same class does not reach 10 %. These differences diminish slightly in the 15–17 yr class. In not

34

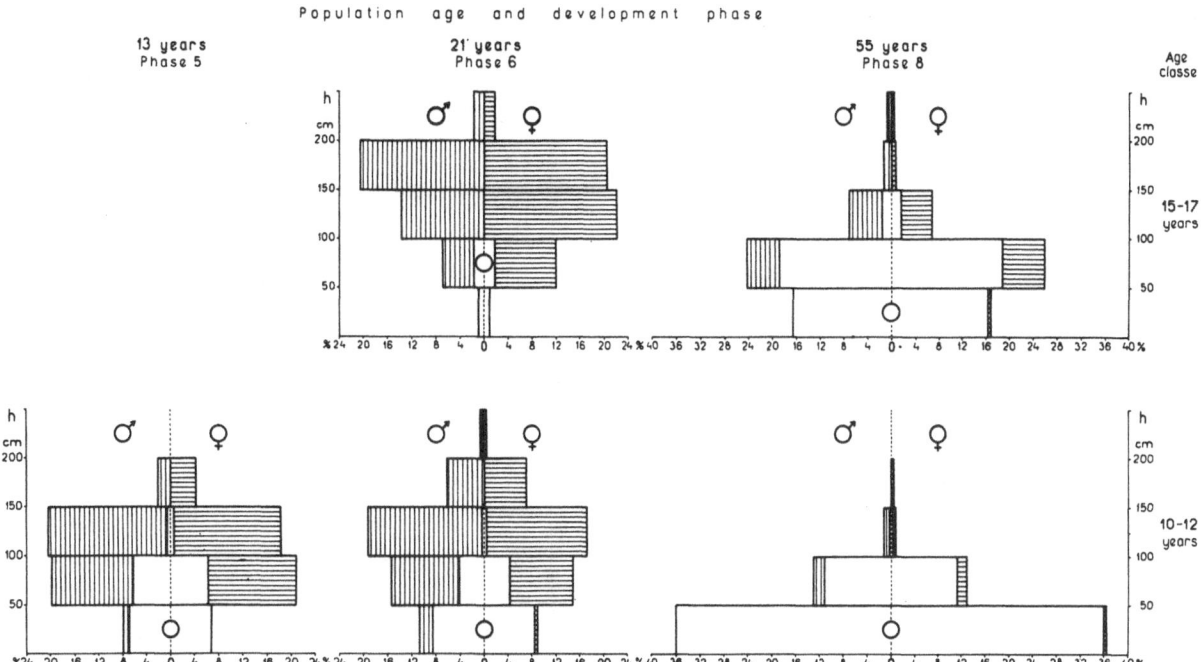

Fig. 13. Sex structure and individual size structure in analogous age classes from the younger, older and the oldest populations of *Juniperus communis*. Detailed analysis of the material is shown in Fig. 12.

overcondensed populations practically all individuals of this age reproduce (see Fig. 7). The height pyramids elaborated for these population classes at various ages demonstrate a distinct growth inhibition of individuals in the older overcondensed population (Figs. 6ABC and 14). The majority of individuals in the overcondensed population does not reach a height of 1 m, whereas in not so dense populations more than one half of the individuals of the same age belong to the height class 1–2 m.

The influence of density on the sex size structure may, therefore, be defined as inhibition of growth of individuals and retardation of the whole age class in reaching the phase of reproduction.

**Conclusions**

It was found that in the course of the succession the *Juniperus* populations change as to the following characteristics: (Fig. 14):

Spatial structure: From random structure in the initial stage to agglomeration structure in the optimal and terminal stages;

Abundance: Type of growth: two-peak left-biassed curve with first peak at optimal stage and second one at terminal stage (Fig. 14);

Age structure: From one-generation initial stage to a multigeneration one beginning at the optimal stage. In the differentiated multi-generation structure the consequences of a 2- or 3 yr cycle in diaspore production are visible (Figs. 6 and 7).

Sex structure:
– participation of identified individuals/involved in generative reproduction/ A relative increase of the participation of identified individuals up to $\frac{2}{3}$ of the population composition in the optimal stage, falling to about $\frac{1}{3}$ in the terminal stage (Figs. 14);

– sex ratio from $\male : \female = 1 : 0.4$ in initial stage through $\male : \female = 1 : 1$ in optimal stage to $\male : \female = 1 : 1.1$ in terminal stage;/ from prevalence of male individuals, via equilibrium to prevalence of female ones (Fig. 14, Table 1);

35

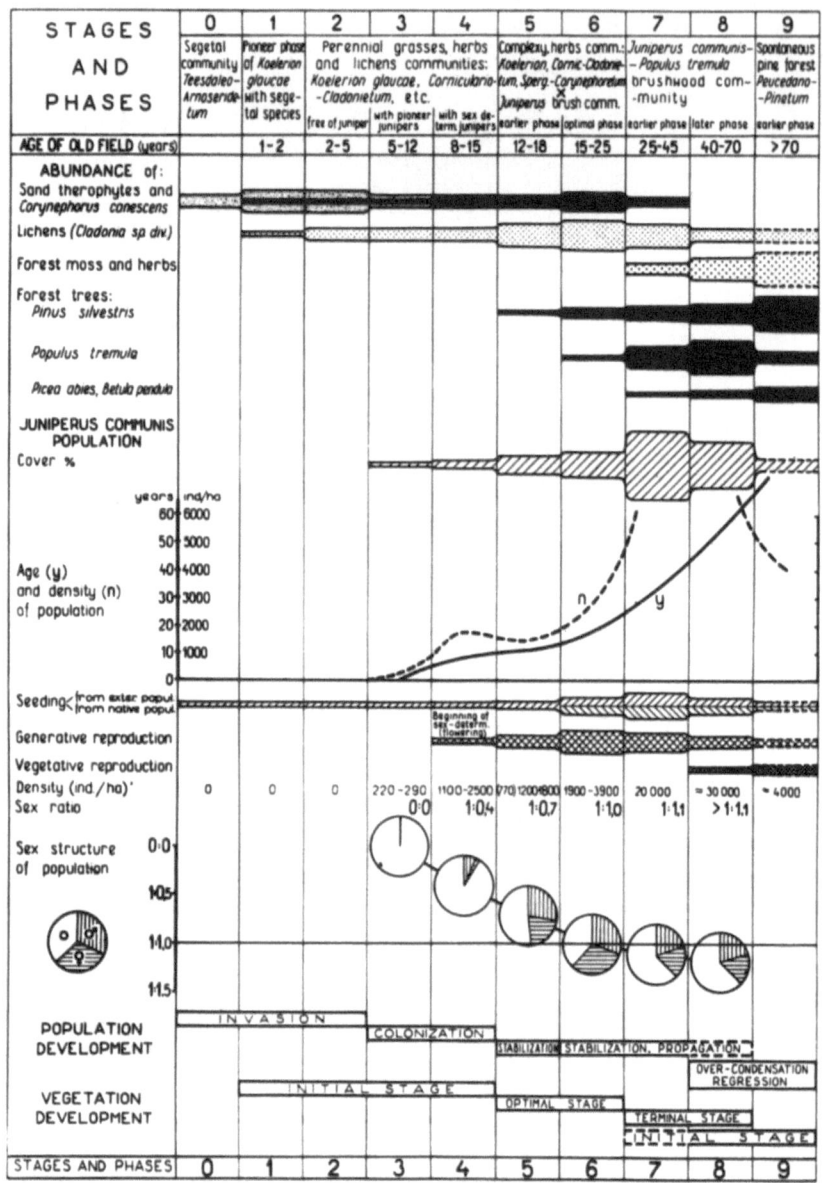

Fig. 14. Plant succession in old fields and development of the population of *Juniperus communis*. At the bottom: The synchronisation of the phases of juniper population development with the phases of vegetation succession.

Types of reproduction: Generative in the whole succession series; exceptionally vegetative, only when imposed by overcondensation or destructuion of erect shoots – only in terminal stage,

Population growth: Substitution of seeding from outside, specific for initial stage, by autochthonous propagation in optimal and terminal stages (Fig. 14);

Diaspore production:
– total of population — Continuous increase from the end of the initial stage to the terminal stage;

– individual — increase to end of initial stage, of optimal stage and decrease in the terminal stage;

36

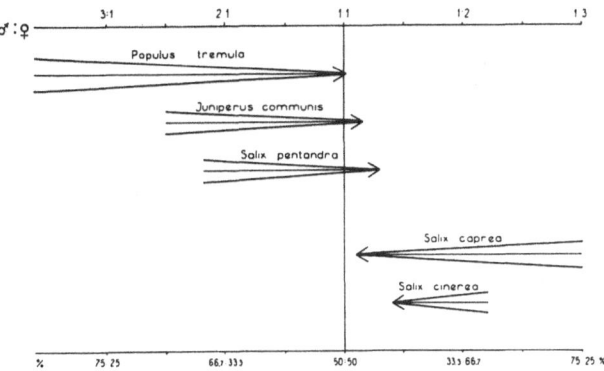

Fig. 15. Supposed sex ratio changes in the populations of pioneer woody species in the vegetation succession.

| Calendar age of individuals at the moment of starting generative reproduction | In the optimal stage of succession at the age of 8-9 yr, in terminal stage retarded and occurring over a period from 10 to 20 (26) yr (Figs. 6 and 14). |
|---|---|

In view of the changes in sex ratio in the population in connection with vegetation dynamics, at least two groups can be distinguished among the pioneer woody species (Fig. 15):

(1) *Juniperus communis*, *Salix pentandra* and eventually *Populus tremula* in which the sex ratio changes from prevalence of male individuals in the early phases of population and vegetation development to an equilibrium in the phase of stabilisation and to a greater or smaller numerical dominance of female individuals in the latest phases;

(2) *Salix caprea*, *S. aurita*, *S. cinerea* and *S. rosmarinifolia*, and their hybrids, the populations of which are characterized by a constant prevalence of female individuals, but the sex ratio of which in the early phases of population and vegetation development reveals a higher contribution of male individuals.

A possible confirmation of the hypothesis of 'patchiness' of the spatial population structure of *Populus tremula* could lead to the distinction of a third group of species. Otherwise the sex structure of the population of this pioneer tree would have to be considered on a much larger scale. Then, however, elimination of additional factors affecting the sex structure of the exceptional role of dioecious woody species in the early phases of succession suggests that their success in colonisation of new habitats is based on the ability to apply colonization tactics and a reproductive strategy exceptional in plants. This possibility derives partly from the different ecological requirements of male and female individuals, an earlier initiation of reproduction by male individuals, and also probably their earlier colonization of pioneer habitats or, at least, their higher adaptability to extreme conditions.

**Summary**

In connection with vegetation succession in old farmland and meadows, changes in the sex structure were analysed, particularly the sex ratio in populations, of *Juniperus communis*, *Populus tremula*, *Salix aurita*, *S. caprea*, *S. rosmarinifolia*, *S. pentandra* etc. Particular attention was given to the changes in the characteristics of *Juniperus communis* populations, associated with succession in old fields, leading from the segetal community *Teesdaleo-Arnoseridetum*, through the therophyte communities with *Corynephorus canescens*, lichens *Corniculario-Cladonietum mitis*, sand sward with the alliance *Koelerion glaucae* to *Juniperus-Populus* brushwood and *Peucedano-Pinetum typicum* forest. The particular succession phases over 70 yr and the phases of development of the pioneer populations of tree species were dated by the dendro-chronological methods.

It was found that in the course of the succession the sex structure, and in particular the sex ratio, changes in the populations of the above-mentioned species. Among the factors that affect the natural populations, the increase in density of the *Juniperus communis* population itself together with changes in the structure of the whole associated community played the most important role. This factor inhibits growth of juniper individuals very strongly and retards the beginning of the generative reproduction phase of a given population by 10–15 yr.

A hypothesis is advanced concerning the role of dioecism in pioneer tree species in the process of colonisation.

In view of the changes in sex ratio in the population in connection with vegetation dynamics, at least two groups can be distinguished among the pioneer woody species:

(1) *Juniperus communis* and *Salix pentandra* in which the sex ratio changes from prevalence of male individuals in the early phases of population (and vegetation) development to an equilibrium in the phase of stabilisation and to a greater or smaller numerical dominance of female individuals /in the latest phases;

(2) *Salix caprea*, *S. aurita*, *S. cinerea* and *Salix rosmarinifolia* and their hybrids, the populations of which are characterised by a constant prevalence of female indivi-

duals, but the sex ratio in the early phases of population (and vegetation) development reveals a higher contribution of male individuals.

## References

Białobok S. 1973. Zagadnienia genetyczne i hodowla. Topole – Populus L. PWN Warszawa – Poznań 12: 315–369.

Blackburn W.H., P.T. Tueller. 1970. Pinyou and Juniper invasion in black sagebrush communities in East-Central Nevada. Ecology 51: 841–848.

Bråkenhielm S. 1977. Vegetation dynamics of afforested farmland in a district of Southeastern Sweden. – Acta Phytogeogr. Suec. 63: 1–106.

Falińska K. 1979. Modification of plant populations in forest ecosystems and their ecotones. Pol. Ecol. Stud. 5,1: 89–150.

Faliński J.B. 1966. Antropogeniczna roślinność Puszczy Białowieskiej jako wynik synantropizacji naturalnego kompleksu leśnego. Diss. Univ. Vars. 13: 1–256.

Faliński J.B. 1972. Potencjalna roślinność naturalna Wysoczyzny Bielskiej. Mater. Zakł. Fitosoc. Stos. UW. Warszawa – Białowieża. 24: 1–23 + wkł.

Faliński J.B. 1977. Research on vegetation and plant population dynamics conducted by Białowieża Geobotanical Station of the Warsaw University in the Białowieża Primeval Forest and in the environ (1952–1977). Phytocenosis 6: 1–132.

Faliński J.B. (1980 in press). Właściwości biologiczne i ekologiczne pionierskich gatunków drzew i krzewów. Biological and ecological properties of pioneer woody species. Wiad. Ekol. (in press).

Farmer R.E. 1964. Sex ratio and sex related characteristic in Eastern Cottonwood. Silvae Genetica. 13: 597–599.

Freeman D.C., L.G. Klikoff & K.T. Harper 1976. Differential resource utilization by the sexes of dioecious plants. Science 193: 597–599.

Hard G. 1975. Vegetationsdynamik und Verwaldungsprozesse auf den Brachflächen Mitteleuropas. Die Erde. 106: 243–276.

Harper J.L. 1977. Population biology of plants. Academic Press. London, New York, San Francisco, 892 pp.

Meisel K. 1978. Vegetationsentwicklung auf Brachflächen. Ref. v. 3 Symposion (13-17.9.1976). Synanthrope Flora und Vegetation Acta Bot. Slov. A.3: 311–318.

Muhle Larsen C. 1970. Recent advances in poplar breeding. Interv. Rev. Forestry Res. 3: 1–67.Academic Press, New York-London.

Pauley S.S., & G.T. Mennel 1957. Sex ratio and hermaphroditism in a natural population of qualing aspen (Populus tremuloides). Minn. For. Not. 55.

Putwain P.D., & J.L. Harper 1972. Studies in the dynamics of plant populations. V. Mechanisms governing the sex ratio in Rumex acetosa and R. acetosella L., J. Ecol. 60: 113–129.

Strzelecki W., & R. Sobczak 1972. Zalesianie nieużytków i gruntów trudnych do odnowienia. PWRiL. Warszawa, 352 pp.

Tüxen R. 1973. Zum Birken-Anflug im Naturschutzpark Lüneburger Heide. Eine pflanzensoziologische Betrachtung. Mitt Flor.-Soz. Arbeitsgem. NF 15/16: 203–209.

Vernet P. 1971. La proportion des sexes chez Asparagus acutifolius L. Bull. Soc. Bot. Fr. 118: 345–358.

Zarzycki K. 1975. Competition between male and female plants of Salix purpurea L. in single and bothsex cultures in different densities. – Abstr. XII Intern. Bot. Congress, July 3–10, 1975 Leningrad p. 175.

Zarzycki K. & L. Rychlewski 1972. Sex ratios in Polish natural populations and in seedlings samples of Rumex acetosa L. and R. thyrsiflorus Fing.-Acta Biol. Crac. Ser. Bot., 15: 135–151.

Accepted 10 December 1979

# SUCCESSION PATTERNS ON MOUNTAIN PASTURES*

G. SPATZ

Lehrstuhl für Grünland lehre, Technische Universität München D 8050 Weihenstephan, W. Germany

Keywords:
Gradient, Mountain, Pasture, Pattern, Succession

## Introduction

Land use is troublesome and seldom pays in mountain regions. Large areas of formerly grazed grasslands were abandoned in the past. Pasture vegetation changes rapidly as soon as cattle, the acting agent, had disappeared.

## Material and methods

The influence of management changes on the vegetation of alpine pastures was studied in the Valley of 'Gastein' in the Austrian Alps. Four subdistricts within the central alpine mountains 'Hohe Tauern' were investigated. Succession was studied from an altitude of 1700 meters up to 2500 m above sea level. 'Stand-structure analysis' (Mueller-Dombois & Ellenberg 1974); 'point quadrat' method (Levy & Madden 1933) and direct gradient analysis (Whittaker 1967, Spatz 1975) were applied.

## Results

How much the vegetation of the studied area was influenced by man in the past can be shown by analyzing the vegetation along a soil nutrition gradient.

Fig. 1 reflects the situation of a gradient beginning immediately below a stable, where soil nutrition is plentiful, and ending in a dwarf shrub heath covering a very poor and highly leached soil. The gradient extends no more than 90 meters.

Fig. 1. Above ground phytomass of species along a

* Nomenclature follows:
F. Ehrendorfer. 1973. Liste der Gefaszpflanzen Mitteleuropas. 2. erw. Aufl.

nutrition gradient (Zitterauer-Alm, elevation 2000 m).

Below the hut *Rumex alpinus* forms a highly productive and dense thicket suppressing almost any other plant species. In the still eutrophic portion some 20 m away from the source of nutrients *Deschampsia cespitosa* and *Ranunculus aconitifolius* achieve a high biomass. At a distance of 60 m *Nardus stricta* and *Festuca rubra*, adapted to poor soil conditions, occur and soon dominate. Finally at a distance of some 80 m, nutrient intake has stopped, and according to very poor soil conditions a dwarf shrub heath developes. The species diversity is very low at the eutrophic side of the gradient, it increases at the middle part, and declines again at the distrophic end. It is obvious that the patterns of succession are highly dependent on the kind of existing vegetation. Some possibilities are being illustrated in the following text.

a) *Succession on a formerly mown meadow with good soil nutrition* (1900 m a.s.l.)

The mowing of a meadow is an anthropogenic factor which keeps the vegetation in a labile balance. As soon as mowing is abandoned, the species composition of the meadow rapidly changes. The situation was analysed in a meadow dominated by tall grasses and *Alnus viridis* saplings. This meadow was mown for the last time about four years ago . The *Alnus* saplings were counted and measured on an area of 250 m². The result of the structural analysis is shown in Fig. 4. An almost even aged stand of alders has been growing up.

Fig. 2. A stand of *Alnus viridis* of almost the same age has grown up and will form a dense shrub within a few years (Zitterauer-Alm 1900 m a.s.l.)

Forest trees are not able to invade the place before the alder stand, which consists of trees of the same age, will

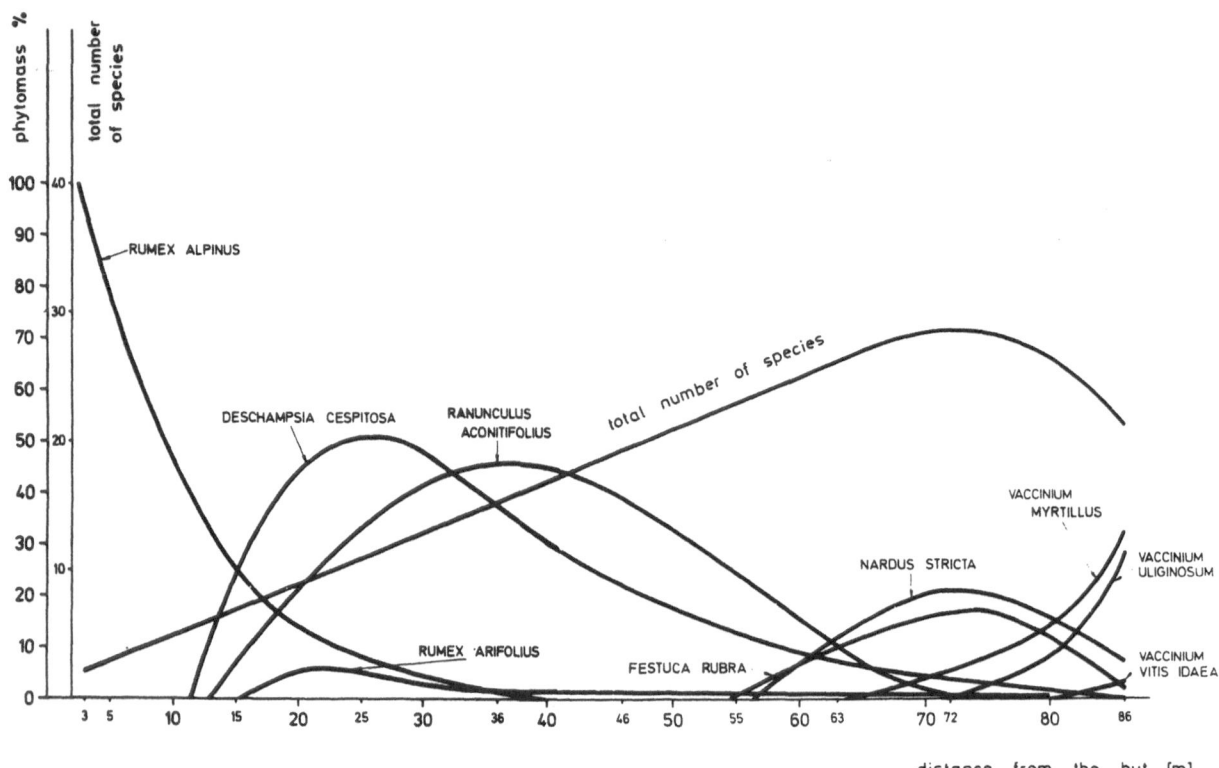

Fig. 1. Above ground phytomass of species along a nutrition gradient (Zitterbauer-Alm. elevation 2000m)

become senile and breaks down. Körner and Hilscher found out in the same area, that alders become senile at an age of 60 years an then suddenly die.

b) *Succession on a secondary dwarf shrub heath below the tree line* (1900 m a.s.l.)

A secondary dwarf shrub heath which was grazed only occasionally in the past was analysed too. This vegetation type has developed from a *Piceetum subalpinum* during

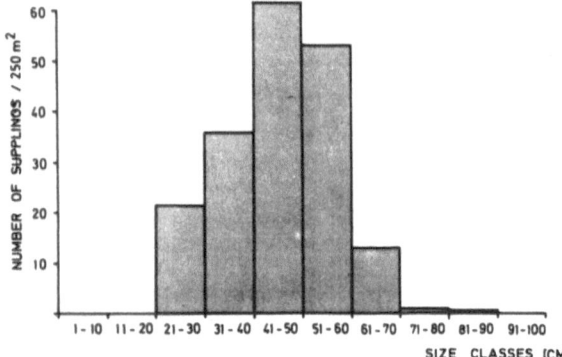

Fig. 2. A stand of *Alnus viridis* of almost the same age has grown up and will form a dense shrub within a few years.

the years of extensive grazing. At the time of analysis it was obvious that succession towards a forest, had already started (Fig. 3).

Fig. 3. Results of a structural analysis of a formerly extensive grazed secondary dwarf shrub heath (1760 m a.s.l.)

*Picea abies*, the main climax tree, and *Sorbus aucuparia* are represented mainly in the lower size classes. They obviously invade this instabile vegetation type. *Alnus viridis* has become senilein this stage of succession. Only a few specimens in the higher size classes are still alive.

c) *Succession in a dwarf shrub heath with Rhododendron ferrugineum near the timber-line* (2200 m a.s.l.)

Near the tree line dwarf shrubs with *Rhododendron ferrugineum* and *Pinus mugo* form the climax vegetation together with the trees *Pinus cembra* and *Larix decidua*. *Picea abies* is close to the boundary of the area where it can survive. *Pinus cembra* and *Larix decidua* trees were suppressed in the past by anthropogenic factors like grazing and burning. As grazing was only occasional in the past, this vegetation type was little modified. Since pasturage was given up little has changed. As structural

40

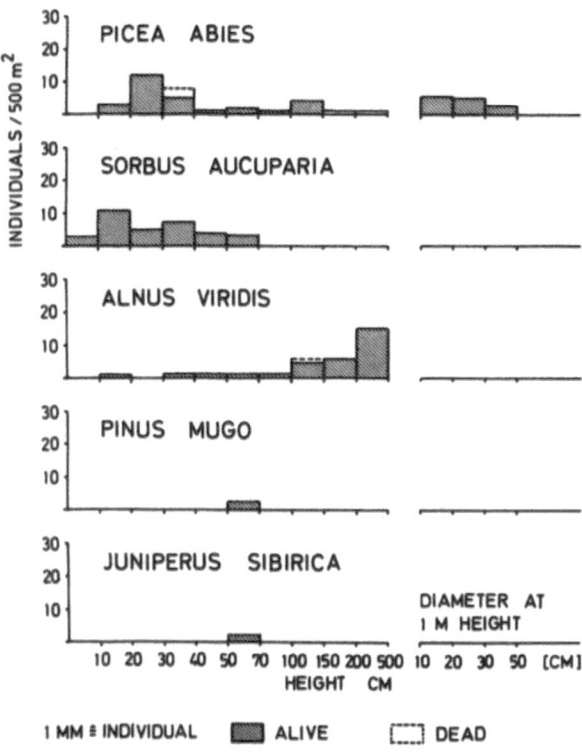

Fig. 3. Results of a structural analysis of a formerly extensive grazed secondary dwarf shrub heath (elevation 1760 m)

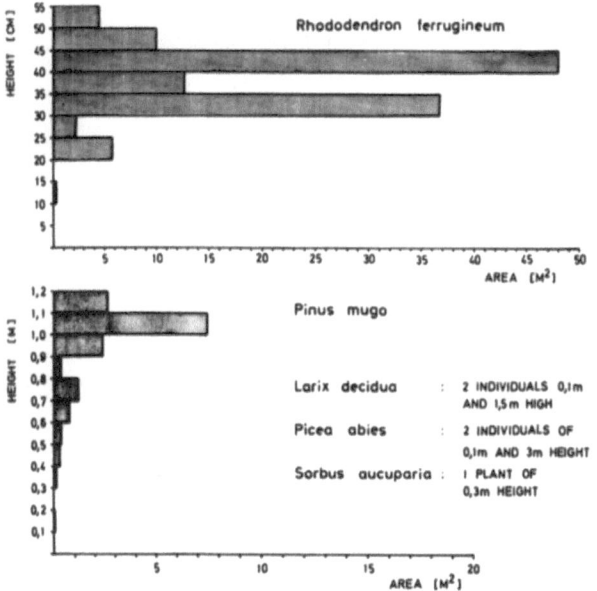

Fig. 4. Dwarf shrub heath near the treeline (2200 m elevation). *Rhododendron ferrugineum* and *Pinus mugo* are distributed normally over the size classes

analysis (Fig. 4) shows, *Pinus mugo* and *Rhododendron ferrugineum* are normally distributed over the size classes. It may be expected that *Larix decidua* will spread in the future but very slowly. In the investigated area of 500 m² only two small trees were found. Succession is quite slow in this relatively stable vegetation and no problem at all.

Above the tree line in the alpine region grasslands were modified only a little by grazing in the past. They redevelop to natural grasslands after cattle has gone.

## Summary

Some successional patterns on mountain pastures in the Hohe Tauern mountains, Austria, are described. A close relation with former pasture management is shown. Very clear gradients in the nutritional status of the soil exist from stables towards more remote areas where no nutrients are added. Dwarf shrub or Alnus viridis woodland vegetation will develop after abandoning pastures, depending on elevation and nutrient status. Above the tree line culturally influenced pastures will rapidly develop to natural grasslands.

## References

Körner, Ch. & H. Hilscher. 1978. Wachstumsdynamik von Grünerlen auf ehemaligen Almflächen an der Zentralalpinen Waldgrenze der Hohen Tavern. In: Veröff. des Österreichischen Masz-Hochgeb. Progr. Hohe Tavern 2, Innsbruck.

Levy, E.E. & E.A. Madden. 1933. The point method of pasture analysis. New Zealand Agric. J. 46: 267–279.

Mueller-Dombois, D. & H. Ellenberg. 1974. Aims and methods of vegetation ecology. I. Willey & Sons, New York, London, Sydney, Toronto, XX + 547 pp.

Spatz, G. 1975. Die direkte Gradientenanalyse in der Vegetationskunde. Angew. Botanik 49: 209–221.

Whittaker, R.H. 1967. Gradient analysis of vegetation. Biol. Rev. 42: 207–264.

Accepted 6 November 1979

# PATTERNS OF PLANT SPECIES DIVERSITY IN FYNBOS VEGETATION, SOUTH AFRICA*

B. M. CAMPBELL[1]** & F. VAN DER MEULEN[2]

[1]Botanical Research Unit, Box 471, Stellenbosch 7600, South Africa
[2]Eemdijk 65, 3754 ND Eemdijk, The Netherlands

Keywords:
Cape, Fire, Mediterranean ecosystem, Sclerophyllous shrubland, Species diversity, Succession

## Introduction

Fynbos is the broad category of sclerophyllous shrublands which dominate the vegetation of the southwestern Cape, South Africa (see Kruger 1979 & Taylor 1978 for detailed descriptions). Plant species diversity both at the landscape level (gamma diversity) and at the community level (alpha diversity) is, supposedly, exceptionally high. Kruger & Taylor (1980) suggest that gamma diversity is apparently much higher than in any of the other world's biogeographic zones, except the tropical rain forest zones and other South African sites of about 1 km². Kruger (1977) states that fynbos communities are probably the richest shrubland communities in the world. He states further that with the exception of plant communities in extreme habitats (for example, where soils are alternatively waterlogged and droughted), alpha diversity is consistently high and some record figures are 83 species on 50 m² and 121 species on 100 m² (data from Taylor 1977, 1978). These figures exclude annuals and geophytes. Kruger et al (1977) suggest that 'it could be that fynbos community diversity is exceeded only by that of tropical rain forests, if at all".

A major goal of ecologists working towards a theory of community organization is to understand the causes of patterns of plant species diversity (Hutchinson 1959, Whittaker 1972, Menge & Sutherland 1976, Peet 1978). There are only a few references to plant species diversity in fynbos, despite its supposed richness. In view of the present interest in plant species diversity patterns we studied the alpha (within-habitat) diversity and the beta (between-habitat) diversity of a small area of mountain fynbos (sensu Taylor 1978) in relation to altitude and fire. These two factors appear to be of major importance in determining fynbos physiognomy.

## Methods

### Field data

The study was carried out at a mountain slope at Jonkershoek, near Cape Town, South Africa. This area is situated in the mediterranean climatic zone of the western Cape, with cool, wet winters and hot, dry summers.

Ten 10 × 5 m plots were subjectively laid out in homogeneous stands of vegetation along an altitudinal gradient (4 plots at 820 m and 6 plots at 400 m). At each altitude, half the plots were located in old vegetation (36 years old) and half the plots in young vegetation (4 years old) (Fig. 1). The altitude gradient reflects undetermined environmental changes in precipitation, temperatures, soil moisture, etc. Soil depth decreases with increasing altitude. Soils are mostly granite-derived but the four high altitude plots have soils that are also partly derived from sandstone colluvium. Aspect (south to southwest facing) and slope (20–25 °) are similar in all plots. In each plot total species composition was determined and the percentage projected canopy cover of each species was estimated. Only vascular plants that are above-ground perennials have been included. The term 'graminoid' has been used to include the usual graminoid plants (*Poaceae*, *Restionaceae* and *Cyperaceae*) as well as some other species with grass-like leaves, e.g. *Bobartia indica* (*Iridaceae*), *Corymbium* sp. (*Astera-*

* Nomenclature follows that used in the Botanical Research Unit Herbarium, Stellenbosch.
** We thank F.J. Kruger, W. Bond, H.P. Linder and J. Somerville for making criticisms on the manuscript.

*ceac*), *Aristea* spp. (*Iridaceae*) and *Schizaea pectinata* (*Schizaeaceae*). These latter species, except for *Bobartia indica*, never contribute more than 1 % cover.

### Data analysis

Diversity relations have been investigated for total species (i.e. vascular plants that are above-ground perennials) and for the woody and graminoid components. These two components contribute more than 90 % of the cover. Diversity relations for the remaining components (ferns, perennial forbs; 5 species) have not been investigated. To calculate importance values of species we used percentage projected canopy cover. Alpha diversity was measured as richness-diversity and as heterogeneity-diversity (sensu Peet 1974). Richness ($S$) was measured as the number of species per 5 m × 10 m plot and heterogeneity-diversity (which combines both richness and evenness) was calculated using the Shannon-Wiener index ($H'$, Whittaker 1972) and Hill's (1973a) second order diversity number ($N_2$; Peet 1974, 1978). $N_2$ is strongly influenced by the importance values of the most important species and is considered to be less effective for inferring diversity than $H'$ (Whittaker 1972). Simpson's index, of which $N_2$ is the reciprocal, is primarily a measure of dominance as degree of concentration of importance values in one or a few species (Whittaker 1972, Peet 1974, 1978). When discussing trends in alpha diversity we are generally discussing trends that are shown by all three measures. $N_2$ trends sometimes deviate from those shown by $S$ and $H'$.

Beta diversity, the extent of differentiation of communities along environmental gradients (Whittaker 1972), has been inferred from similarity coefficients. Similarity values were calculated using three techniques: (1) log transformed importance values and the Czekanowski coefficient, (2) untransformed importance values and the Czekanowski coefficient, and (3) Coetzee and Werger transformed importance values and the relativized Czekanowski coefficient. The choice of these techniques is based on the work of Campbell (1978). All three techniques gave similar results and only the results using technique (i) have been reported here. Ordinations (reciprocal averaging, Hill 1973b) and numerical classifications (group-average sorting, Campbell 1978) were produced and used to infer beta diversity relations. The environmental factors that could be related to the major groups of a classification or ordination were considered to be the important factors contributing to a high beta diversity. The ordinations and classifications added little to what has been presented in Fig. 2, and thus are not reported here.

### The vegetation types

The six low altitude plots are from the proteoid zone and the four high altitude plots are from the restioid-ericoid zone (sensu Taylor 1978) (Fig. 1). The physiognomic features of the vegetation types and some of the predominant species are presented in Fig. 1. The low altitude old fynbos is a tall scrub, dominated by the proteoid shrub *Protea neriifolia*, the narrow-leaved shrub *Brunia nodiflora* (in places), the cupressoid tree *Widdringtonia cuppressoides*, and the tall restioid *Cannamois virgata*. The low altitude young vegetation is mostly graminoid with *Cannamois virgata*, the grass *Merxmuellera rufa* and the evergreen geophyte *Bobartia indica* being the predominant species. The high altitude old vegetation is mostly graminoid and is dominated by the restioids *Hypodiscus albo-aristatus* and *Elegia juncea*, the grass-like Cyperaceae *Tetraria involucrata*, and *Bobartia indica*. The high altitude young vegetation is also mostly graminoid; dominants are *Hypodiscus albo-aristatus*, *Tetraria involucrata*, *Bobartia indica* and the ericoid shrub *Erica pinea*. Typical for the

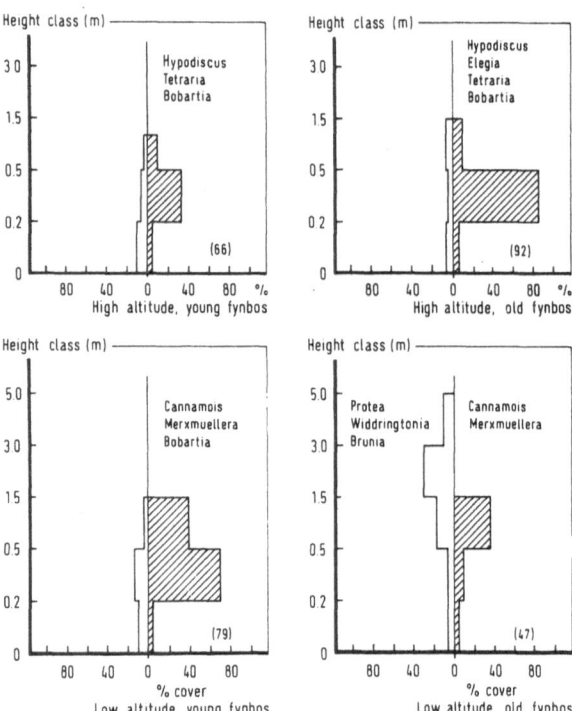

Fig. 1. Average cover distribution of woody species (not hatched) and graminoid species (hatched) among height classes in old and young fynbos at different altitudes. Numbers in brackets indicate cover of graminoid component as a % of the total cover. Genus names of predominant species are given.

low altitude unburned fynbos is the dead standing material (mainly *Protea neriifolia*), contributing up to 15 % cover, and a fairly thick (about 5 cm) layer of plant litter covering half the plot or more.

## Diversity

### Alpha diversity

Table 1 gives the results for alpha diversity ($S$, $H'$ and $N_2$) for total species and for graminoid and woody species. Total species diversity according to all indices is lower in the old fynbos than in the young fynbos at all altitudes. This reduction of diversity with age is due to the process of senescence in fynbos. Post-fire succession in fynbos has the following characteristics (Kruger 1977). All or most of the pre-fire species regenerate within the first twelve months after fire. Very few species enter the community after this period. In the absence of further fire the species eventually reach their maximum life span and senescence then die. Thus, post-fire succession is marked by a gradual decrease in species richness after the immediate post-fire phase. The old fynbos in this study can be regarded as senescent (sensu Kruger 1977).

Changes in alpha diversity are largely a reflection of changes in diversity of the lower strata – the herbaceous and low shrub strata. The presence of a tall shrub and restioid strata is correlated with a reduction in cover of the lower strata (cf. Fig. 1) and a reduction in diversity. Thus the low altitude old vegetation with its tall shrub and restioid strata has the lowest alpha diversity. In the high altitude old vegetation which lacks a tall stratum there is a higher cover of graminoids and low shrubs, and a higher alpha diversity. This overstory/understory and diversity relationship has also been indicated for Australian heathlands (Specht 1979). Similarly, Auclair & Golff (1971), del Moral (1972), Whittaker (1960, 1965, 1972), Glenn-Lewin (1977), Peet (1978) and Campbell (1979) have noted that dense tree canopies reduce herb richness, that the herb strata are the primary determinants of total alpha diversity and that the reduction of herb diversity is correlated with a reduction in herb cover.

A major constituent of lower strata (less than 1 m) in fynbos is the graminoid component (Fig. 1). Thus a reduction in diversity due to overtopping by taller strata should be reflected in a reduction of diversity of the graminoid component. These patterns are indeed the case where this overtopping occurs, namely on the gradients from short to tall fynbos (i.e. at low altitudes from young to old fynbos and in old fynbos from high to low altitudes; cf. Fig. 1). The woody component shows little pattern in alpha diversity perhaps because it is a heterogeneous group of species, including tree seedlings that migrate into the community late in succession, tall shrubs, dwarf creeping shrubs, etc. Perhaps the only pattern in alpha diversity of

Table 1. Alpha diversities in fynbos, expressed as species richness ($S$) and heterogeneity-diversity ($H'$ = Shannon-Wiener index, $N_2$ = reciprocal of Simpson's index).

|  | Plot no. | Total species | | | Woody species | | | Graminoid species | | |
|---|---|---|---|---|---|---|---|---|---|---|
|  |  | S | H' | $N_2$ | S | H' | $N_2$ | S | H' | $N_2$ |
| High altitude, old fynbos | 9 | 29 | .86 | 4.5 | 12 | .72 | 4.0 | 15 | .69 | 3.4 |
|  | 11 | 27 | .89 | 5.5 | 14 | .53 | 2.0 | 13 | .79 | 4.7 |
| young fynbos | 10 | 39 | 1.06 | 9.0 | 21 | .68 | 3.2 | 17 | .79 | 4.7 |
|  | 12 | 35 | .90 | 5.5 | 15 | .51 | 2.3 | 17 | .66 | 3.1 |
| Low altitude, old fynbos | 1 | 23 | .85 | 5.3 | 15 | .64 | 3.4 | 5 | .40 | 1.8 |
|  | 3 | 25 | .84 | 5.3 | 16 | .68 | 3.4 | 6 | .42 | 2.1 |
|  | 5 | 24 | .61 | 3.0 | 16 | .45 | 2.1 | 6 | .21 | 1.2 |
| young fynbos | 2 | 30 | .93 | 7.1 | 16 | .63 | 3.4 | 12 | .64 | 3.8 |
|  | 4 | 33 | 1.01 | 7.6 | 17 | .76 | 4.5 | 13 | .74 | 4.5 |
|  | 6 | 34 | .95 | 6.6 | 14 | .60 | 2.6 | 17 | .69 | 3.7 |

woody species is the decrease in diversity from young to old vegetation at high altitudes. This decrease is correlated with a decrease in cover of woody species relative to the graminoid species (Fig. 1). It would seem that graminoid species are able to outlive woody species, at least in situations when overtopping does not occur.

*Beta diversity*

Fig. 2a indicates that a higher beta diversity is associated more with the altitude gradient than with the temporal gradient (age after fire).

That the temporal gradient has relatively low beta diversity is not surprising considering the nature of succession in fynbos. If fynbos succession consisted of an orderly and predictable series of species replacements in time, then one would expect a higher beta diversity along the temporal gradient. In fynbos, however, succession fits Egler's (1954) 'initial floristic composition' model of succession. All or most of the species of the succession are present at the site within a year of the fire and succession represents a process of change reflecting the differential life-cycles and growth forms of the initial species (Kruger 1977). The change that occurs on the temporal gradient is mostly quantitative, not qualitative and thus beta diversity along the temporal gradient is relatively low.

The four year old fynbos in our study is on a fire-break and is burned every eight years. This short burning cycle has probably caused the local extinction of many of the seed-regenerating species. Thus, the beta diversity recorded on the temporal gradient is probably higher than what would naturally occur on a gradient from four to 36 year old fynbos. That fire is not very important in determining beta diversity, is suggested by the phytosociological study of Werger, Kruger & Taylor (1972) in fynbos of the same catchment area. Stands of widely different structure, presumably due to fire, were placed in the same floristic association because of their similarity in species composition.

On the temporal gradient the beta diversity is higher at low altitudes than at high altitudes and on the altitude gradient the beta diversity is higher in the old fynbos (Fig. 2a). On both the altitude and temporal gradients the higher beta diversity occurs where there is a greater range of structure i.e. on the gradients that go from short fynbos to tall fynbos (Fig. 1). This higher beta diversity on the short to tall fynbos gradients probably occurs because (a) the tall fynbos contains a set of tall species that are absent from the short fynbos. Most should have been present in the short low altitude fynbos but the seed regenerating species have undergone local extinction due to too frequent burning; (b) the tall fynbos is lacking in a set of short shrubs and graminoids because of overtopping. Thus we are suggesting

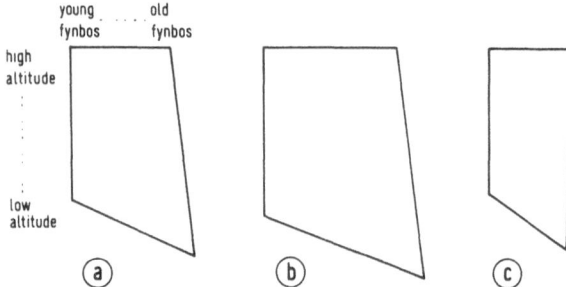

Fig. 2. Schematic presentation of beta diversities between fynbos communities on the altitude (vertical axes) and temporal (horizontal axes) gradients. The length of a line is proportional to the diversity on the gradient. Beta diversity = 100 − (average percentage of similarity between the plots at each end of an axis). a = all species taken into account, b = woody species taken into account, c = graminoid species taken into account.

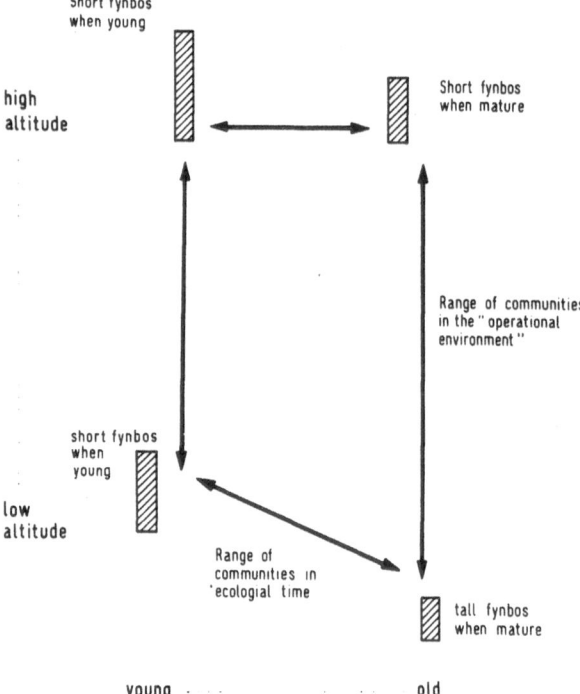

Fig. 3. Model of fynbos diversity patterns. Beta diversities are proportional to the length of the arrows. Alpha diversities are proportional to the height of the vertical bars. 'Operational environment' = the environmental factors that impinge directly on plants (Mason & Langenheim 1957).

that tallness is accentuating floristic differences between communities.

The woody component always has a higher beta diversity than the graminoid component and is the primary determinant of total beta diversity (Fig. 2). This is probably as a result of the woody component being a more heterogenous group of species (i.e. a wider range of growth forms: short shrubs, tall shrubs, coppice vs. seed regenerating, climbers, small trees etc.) than the graminoid component. Higher beta diversity on the temporal gradient at both altitudes of the woody component relative to the graminoid component is especially marked probably due to (a) the absence of relatively short-lived shrubs in the old vegetation (it appears as if graminoids can outlive many shrubs, at least at the high altitude) and (b) the absence of many seed-regenerating species in the young vegetation (most of these seed-regenerating species are shrubs or short trees; the graminoids mostly regenerate vegetatively).

In conclusion, we present our results in a two dimensional graphic model of fynbos diversity (Fig. 3).

## Summary

In fynbos vegetation of the southwestern Cape Province, South Africa, plant species diversity relations (alpha and beta diversity) are studied in a local context of variation across two interacting gradients, a temporal gradient (age after fire) and an altitudinal gradient.

Alpha diversity is lower in the older fynbos at high and low altitudes. This can be related to the nature of succession in fynbos. The lowest alpha diversity occurs in the low altitude old vegetation. This appears to occur because of the decrease of understory cover due to overtopping by tall shrubs and tall restioids.

Beta diversity is due more to the altitude factor than to fire. This can also be related to the nature of fynbos succession. In general the trends shown by alpha and beta diversity can be explained with reference to fynbos succession, the regenerating strategies after fire (seed vs. vegetative) and the effect of the overstory on the understory.

## References

Auclair, A.N. & T.G. Goff. 1971. Diversity relations of upland forests in the western Great Lakes area. Amer. Nat. 105: 499–528.

Campbell, B.M. 1978. Similarity coefficients for classifying relevés. Vegetatio 37: 101–109.

Campbell, B.M. 1979. Some mixed hardwood forest communities of the coastal ranges of southern California. Phytocoenologia (in press).

Egler, F.E. 1954. Vegetation science concepts: 1. Initial floristic composition, a factor in old field vegetation development. Vegetatio 4: 412–417.

Glen-Lewin, D.C. 1977. Species diversity in North American temperate forests. Vegetatio 33: 153–162.

Hill, M.O. 1973a. Diversity and evenness: a unifying notation and its consequences. Ecology 54: 427–432.

Hill, M.O. 1973b. Reciprocal averaging: an eigenvector method of ordination. J. Ecol. 61: 327–249.

Hutchinson, G.E. 1959. Homage to Santa Rosalia, or why are there so many kinds of animals? Amer. Nat. 93: 145–159.

Kruger, F.J. 1977. Ecology of Cape Fynbos in relation to fire. Proc. Symp. environmental consequences of fire and fuel management in Mediterranean ecosystems. USDA For. Serv., Gen. Techn. Rep. WO-3: 230–244.

Kruger, F.J. 1979 South African heathlands. In Specht, R. (ed.) Heathlands and related shrublands of the world. In D.W. Goodall (ed.) Ecosystems of the world. Elsevier (in print).

Kruger, F.J., D.P. Bands, B.J. Durand & R.A. Haynes. 1977. Ecology and management of Cape Fynbos: towards the conservation of a unique biome type. Proc. S.A. Wildlife Management Assoc. 2nd Int. Symp. (Unpublished).

Kruger, F.J. & H.C.Taylor. 1980. Plant species diversity in the Cape fynbos. 1. Gamma and delta diversity. Vegetatio.

Mason, H.L. & J.H. Langenheim. 1957. Language analysis and the concept environment. Ecology 38: 325–340.

Menge, B.A. & J.P. Sutherland. 1976. Species diversity gradients: synthesis of the roles of predation, competition and temporal heterogeneity. Amer. Nat. 110: 351–369.

Moral, R. del. 1972. Diversity patterns in forest vegetation of the Wenatchee Mountains, Washington. Bull. Torrey Bot. Club 99: 57–64.

Peet, R.K. 1974. The measurement of species diversity. Ann. Rev. Ecol. Syst. 5: 285–307.

Peet, R.K. 1978. Forest vegetation of the Colorado Front Range: patterns of species diversity. Vegetatio 37: 65–78.

Specht, R. 1979. The schlerophyllous (heath) vegetation of Australia: The Eastern and Central States. In R.Specht: Heathlands and related shrublands of the world. In D.W. Goodall (ed.) Ecosystems of the world. Elsevier (in print).

Taylor, H.C. 1977. Aspects of the ecology of the Cape of Good Hope Nature Reserve in relation to fire and conservation. Symp. Environmental consequences of fire and fuel management in Mediterranean ecosystems. USDA. For. Serv., Gen. Techn. Rep. WO–3: 483–487.

Taylor, H.C. 1978. Capensis. In M.J.A. Werger, (Ed.), Biogeography and ecology of Southern Africa, p. 171–229. Junk, The Hague.

Werger, M.J.A., F.J. Kruger & H.C. Taylor. 1972. A phytosociological study of the Cape Fynbos and other vegetation at Jonkershoek, Stellenbosch. Bothalia 10: 599–614.

Whittaker, R.H. 1960. Vegetation of the Siskiyou Mountains, Oregon and California. Ecol. Monogr. 30: 279–338.

Whittaker, R.H. 1965. Dominance and diversity in land plant communities. Science 147: 250–260.

Whittaker, R.H. 1972. Evolution and measurement of species diversity. Taxon 21: 213–251.

Accepted 14 December 1979

# DIVERSITY AND STABILITY IN GARRIGUE ECOSYSTEMS AFTER FIRE*

L. TRABAUD[1] & J. LEPART[2]

[1]CEPE/CNRS B.P. 5051, 34 033 Montpellier Cedex, France
[2]Ecothèque méditerranéene, B.P. 5051, 34 033 Montpellier Cedex, France

**Keywords:**

Bas-Languedoc, Diversity, Fire, Fugacity, Garrigue ecosystems, Mediterranean, Southern France, Stability

## Introduction

By its frequent recurrence fire is an important factor for the dynamics of plant communities in the French mediterranean region. Up to now it was considered to create series of more and more degraded successive stages (Braun-Blanquet 1936, Kuhnholtz-Lordat 1938, 1958, Kornas 1958, effect of fire was studied by Le Houerou (1974, 1977) and as due to fire action. But, the processes by which vegetation is recovering after fire were not tackled precisely: the authors only compared different associations.

Elsewhere in the Mediterranean Basin, the ecological of fire was studied by Le Houerou (1974, 1977) and Naveh (1974a), 1974b, 1975, 1977). However, they did not analyse exactly the vegetation succession in the course of time after the fire.

In California, vegetation succession after fire was intensively studied in the chaparral (Sampson 1944, Stone & Juhren 1951, Horton & Kraebel 1955, Sweeney 1956, 1967, Biswell 1963, Patric & Hanes 1964; Hanes & Jones 1967, Hanes 1971, Vogl & Schorr 1972). According to these authors, chaparral is well adapted to withstand recurring fires. Hanes (1971) mentioned 'autosuccession' of chaparral; he considered chaparral as a stable plant community for thousands of years.

The present contribution deals with the following problems: The determination of species development in burnt areas is rather unknown: some species settle after the fire, others were pre-existent and are developing considerably as soon as competition lessens by the help of fire. In fact, what is the antagonism between the species

Nomenclature: P. Fournier (1966) Les quatre flores de France.

of the communities which existed before fire and the foreign species which settle afterwards? Which will finally occupy the burnt area?

Moreover, does fire make the flora more common or richer? Do communities get more dynamical, i. e. is a secondary succession involved, or does fire lead towards a new equilibrium? Are the phenomena the same for all communities? In short, our aim is to know if fire is, or is not, a factor of diversity and stability in garrigue ecosystems.

## Method, device and kind of observations

The 'direct method' (Pavillard 1935), or diachronic method, based on permanent plots was chosen to observe the vegetation recovery in a burnt area; This method allows to follow relatively small both floristic and structural changes.

Our study deals mainly with the first years following fire (10–11 yr); they are of primary importance for the establishment of vegetation.

The 47 studied plots are located in formly burnt areas in Bas-Languedoc, representing the most frequently encountered communities in this area.

- dense *Quercus ilex* coppices, where the tree cover was over 50 % three years after fire;
- open *Quercus ilex* coppices, where the tree cover was always below 50 % during the observation period
- dense *Quercus coccifera* garrigues where the cover was over 90 % three years after fire;
- open *Quercus coccifera* garrigues, where the cover was always below 90 %;
- *Pinus halepensis* woodlands;
- *Rosmarinus officinalis* garrigues

– *Brachypodium ramosum* swards

– *Brachypodium phoenicoïdes* swards

The observation plots were chosen with the help of the 'Service d'Incendie et de Secours du département de l'Hérault', which authorized us to examine all fire reports since 1962. Thus we could draw a map of the burnt areas in the county and date all recorded wildfires.

After a reconnaissance of the burnt areas, the study plots were selected according to the apparent homogeneity of the stands.

Each observation plot consists of a permanent 20 m. long line (Levy & Madden 1933, Long 1957, 1958, Daget & Poissonet 1971). Reference posts are cemented into the ground. Observations are done every 10 cm. Presence and number of hits per species are noted.

In addition to these observations, a floristic survey of species present in a 100 m² plot (in. e. in a 2.5 m wide strip at either side of the line) is done. Only the results dealing with these floristic lists are taken into account in this paper.

During the first five years which followed burning the observations were carried out every year in spring. Afterwards, the vegetation was observed only every two years, since the stands appeared to stabilize.

### Development of floristic composition

The main problem is how the different species of a plant community establish after a fire has totally destroyed the above-ground part of the vegetation. We considered three important points:
– the development of floristic richness,
– the relative stabilization of floristic enrichment,
– the spatial changes in floristic composition.

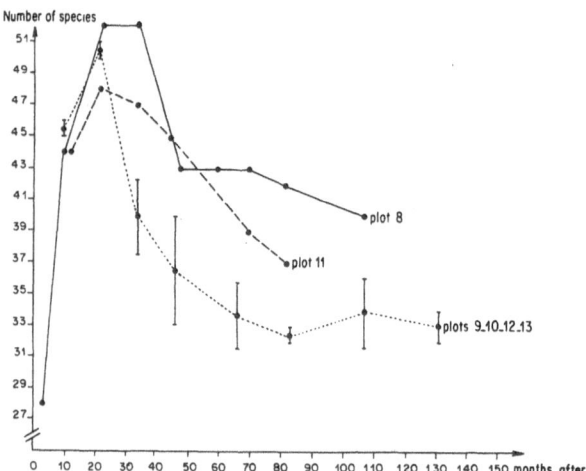

Fig. 2. Floristic richness of open *Quercus ilex* coppices.

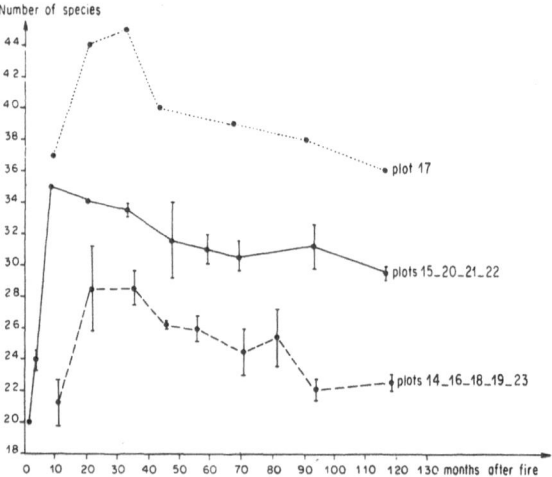

Fig. 3. Floristic richness of dense *Quercus coccifera* garrigues.

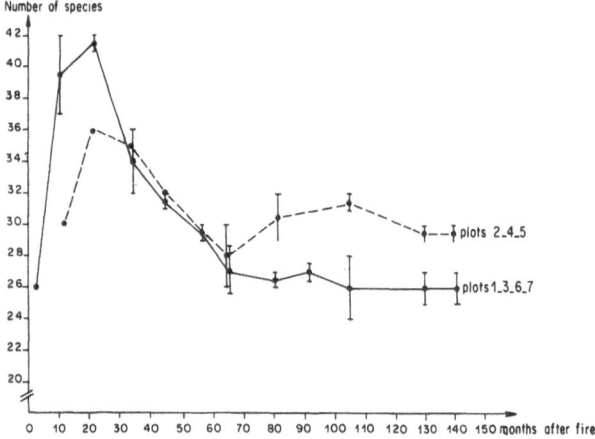

Fig. 1. Floristic richness of dense *Quercus ilex* coppices.

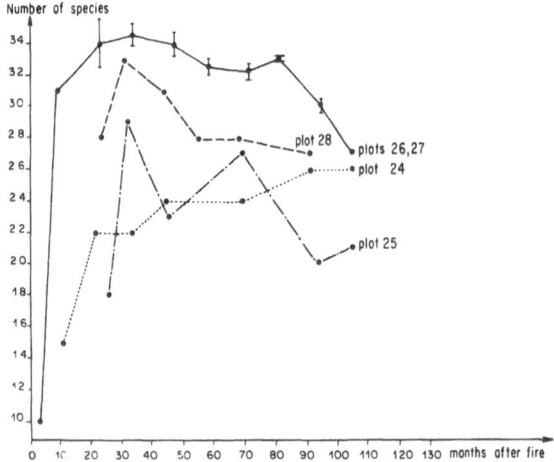

Fig. 4. Floristic richness of open *Quercus coccifera* garrigues.

50

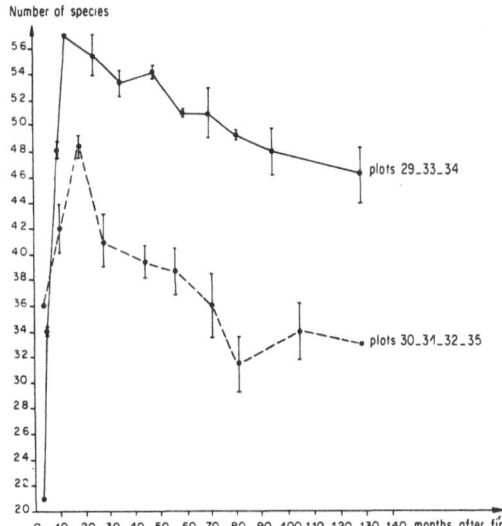

Fig. 5. Floristic richness of *Pinus halepensis* woodlands.

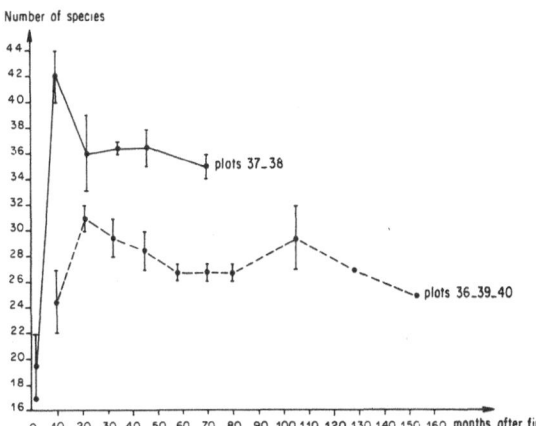

Fig. 6. Floristic richness of *Rosmarinus officinalis* garrigues.

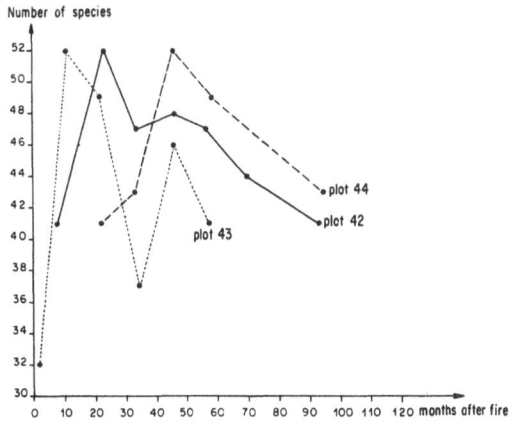

Fig. 7. Floristic richness of *Brachypodium ramosum* swards.

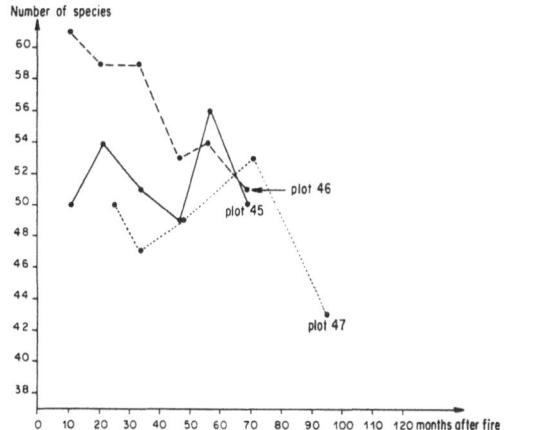

Fig. 8. Floristic richness of *Brachypodium phoenicoides* swards.

## Development of floristic richness

The floristic richness corresponds to the number of taxa encountered in plots at each time of observation.

During the years which follow fire, the development of floristic richness follows a general model. Immediately after a fire, the ground is entirely bare; then species appear progressively. Floristic richness of the communities is low during the first twelve months. It grows gradually to reach a maximum between the 10th and 40th month. Then, the floristic richness diminishes. Finally, a relative stabilization does appear from the 60 th month onwards.

The generally higher number of species during the first three years can be ascribed to the opening of the vegetation cover created by fire, to the disappearence of litter and to the richness in nutrients of the layer of the upper soil (cf. Biswell 1974, Harper 1977). Some species may come from outside the communities; they will disappear again as the stands get older.

To compare the communities, floristic richness can be measured by:
– the maximal richness at any time during the observations
– the richness at the time of the last observation (at the end of the 10th or 12th year).

Floristic richness appears to be linked with the type of community (Table 1 and Figs 1–8). So, *Brachypodium phoenicoïdes* swards do possess the highest floristic richness, followed by *Pinus halepensis* woodlands and *Brachypodium ramosum* swards; whereas *Quercus coccifera* garrigues present the lowest numbers of species.

Floristic richness is high and does not differ much from that of the more mature communities (Bharucha 1932, J. Braun-Blanquet 1935, 1936, G. Braun-Blanquet

51

Table 1. Comparison of floristic richness between the different studied communities.

| | Extreme values of maximal richness | Extreme values of last observation | Classification | Extreme values of floristic richness of corresponding associations | Extreme values of the similarity coefficient between our lists and those of the corresponding associations |
|---|---|---|---|---|---|
| dense *Quercus ilex* coppices | 36 and 42 | 25 and 30 | 6 | 19 and 30 J. BRAUN-BLANQUET (1936) | 0.08 and 0.47 |
| open *Quercus ilex* coppices | 48 and 54 | 32 and 40 | 4 | – | 0.07 and 0.42 |
| dense *Quercus coccifera* garrigues | 21 and 45 | 22 and 36 | 7 | 12 and 33 J. BRAUN-BLANQUET (1935) | 0.16 and 0.60 |
| open *Quercus coccifera* garrigues | 26 and 35 | 21 and 30 | 8 | – | 0.12 and 0.60 |
| *Pinus halepensis* woodlands | 42 and 57 | 32 and 51 | 2 | 26 G. BRAUN-BLANQUET (1936) | |
| *Rosmarinus officinalis* garrigues | 30 and 44 | 25 and 39 | 5 | 35 G. BRAUN-BLANQUET (1936) | |
| *Brachypodium ramosum* swards | 52 | 41 and 43 | 3 | 38 and 62 BHARUCHA (1932) | 0.10 and 0.62 |
| *Brachypodium phoenicoides* swards | 53 and 61 | 43 and 51 | 1 | 32 and 67 SOROCEANU (1936) | 0.04 and 0.28 |

these authors all used 100 m² plots. Jaccard's similarity coefficient was used to improve the floristic comparison (Table 1).

*Limits of the model*

The general model seems to be quite representative for *Quercus ilex* coppices and *Pinus halepensis* woodlands, and most of the *Quercus coccifera* and *Rosmarinus* garrigues. The development of floristic richness of *Brachypodium ramosum* or *B. phoenicoïdes* swards tends to deviate. This could be explained by the proportion of annuals and short-lived low perennials encoutered in those communities (Bharucha 1932, Soroceanu 1936, Braun-Blanquet et al. 1952). As a matter of fact, when only phanerophytes and chamaephytes are considered, the development curves for floristic richness follow a very simply and broadly distributed model: the number of species increases regularly during the first three years, then tends to stabilize.

Although in all the studied communities, annuals are relatively frequent during the first two or three years, it is only in *Brachypodium ramosum* and *Brachypodium phoenicoïdes* swards that they continue to play an important part. Probably their germination is related to factors

we cannot control (amount of viable seeds in the soil type of weather, comings and goings of flocks of sheep).

Another irregularity of the model is due to the time elapsed between the fire date and the date of the first observation. It seems to be difficult to find all the species present in a rather dense vegetation at this first observation.

*Stabilization of the floristic enrichment*

The curves of the development of floristic richness are similar to those which are observed when an unbalanced biological system returns to a metastable state. The stabilization can be described through – the change in fugacity pattern of species, and – the emergence of the 'terminal' community (defined by the last observation)

Fugacity: proposal for an index

A species is called fugacious when it does not remain on the plot all along the period of observations. The fugacity of any species is measured by the number of observations in which it is missing. A fugacity index for the floristic ensemble of a plot should correspond to the mean value of the fugacities of the species present on the plot at a given time.

52

The fugacity index can be considered as a measure of floristic stability of a plot: if the index is high, the community has not reached a stable state, if it equals zero, the community is floristically stable. The fugacity index IF can be computed with the following formula:

$$IF = (F_{max} - \bar{F}_i)/F_{max}$$

Maximal frequency $(F_{max})$ equals the number of observations done on the same plot.

Mean frequency $(\bar{F}_i)$ corresponds to the mean number of observations which possess the species present in the observation i.

The fugacity index can vary between 0 and $(F_{max} - 1)/F_{max}$. It is minimal if all species at one observation are present in the whole series analyzed on the same plot. The community is then said to be in floristical equilibrium. It is maximal if all species observed at observation i are only present at this time. In our study, the fugacity index is not very high; it rarely exceeds 0.25.

The fugacity index is low immediately after the fire and remains low during the first year (Figs. 9–16). The species which are the first to appear on the plot do remain thereafter. Most often, they are species which have strong stumps or rhizomes to withstand fire. Fugacity reaches its maximal value during the second and third year after a fire when floristic richness is at its maximum. Therefore, the richness of the intermediate stages proceeds clearly from species which temporarily add to the community richness and disappear later on. Most often these species are adventitious therophytes. Then the fugacity index decreases progressively to stabilize around 0.10 towards the 8th year after the fire.

Communities which present the highest fugacity during

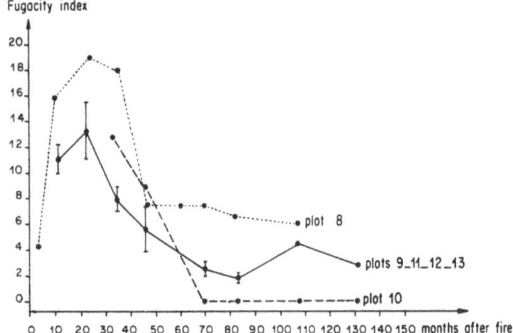

Fig. 10. Development of the fugacity index ($\times$ 100): open *Quercus ilex* coppices

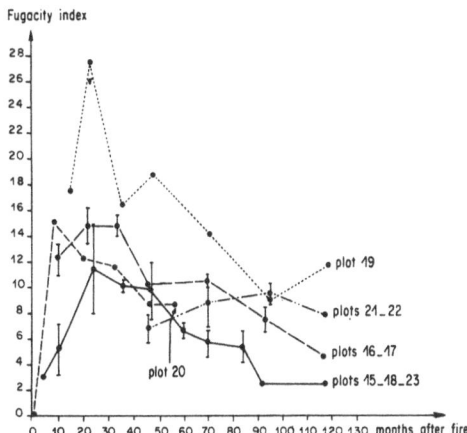

Fig. 11. Development of the fugacity index ($\times$ 100): dense *Quercus coccifera* garrigues.

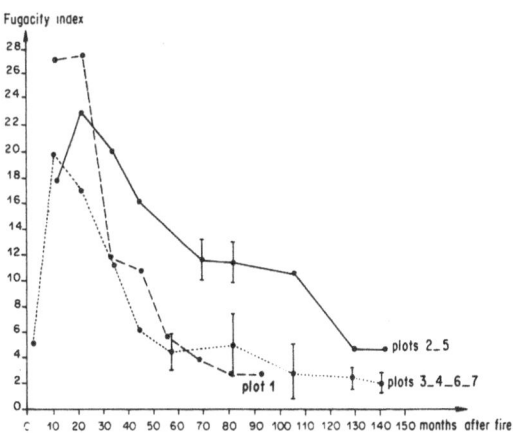

Fig. 9. Development of the fugacity index ($\times$ 100): dense *Quercus ilex* coppices.

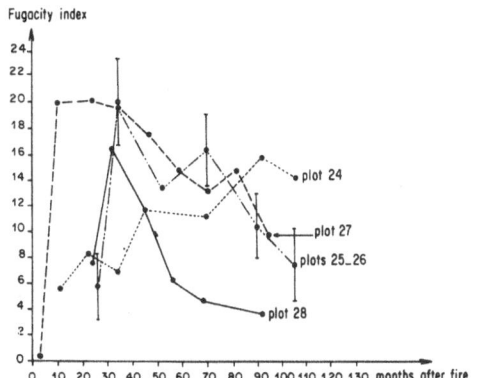

Fig. 12. Development of the fugacity index ($\times$ 100): open *Quercus coccifera* garrigues.

53

the first three years after fire are the dense *Quercus ilex* coppices, *Pinus halepensis* woodlands and *Rosmarinus* garrigues. Whereas those which have the less fugacious species during the same period are dense or open *Quercus coccifera* garrigues and open *Quercus ilex* coppices. This is not surprising for the *Quercus coccifera* garrigues where the vegetation rapidly closes up and prevents any species outside the community to enter.

Emergence of the 'terminal' community

The emergence of the terminal community is determined by the moments the species of the community present at the last observation appeared during the total period of observation. An 'emergence index' can be calculated as the percentage of the species encountered at one observation, which are also found in the terminal community.

Most of the species which belong to the terminal community appear very quickly: one year after the fire 70 % of the plots already possess more than 75 % of the species which are still present at the time of the last observation (Fig. 17), after two years more than 80 % and in five years almost 100 %. The return towards a metastable state, is thus very quick. It follows that there is no real secondary succession, in the sense that species, or communities, do not follow each other on the same plot, after the disturbance of fire.

This phenomenon can be interpreted as an example of resilience as defined by Holling (1973) and Boesch (1974). Although for many authors the notion of resilience is not very different from that of stability, the latter one is lacking precision (Lewontin 1969, Margalef 1969, Holling 1973, Boesch 1974, Golley 1974, Goodman 1974, Orians 1974).

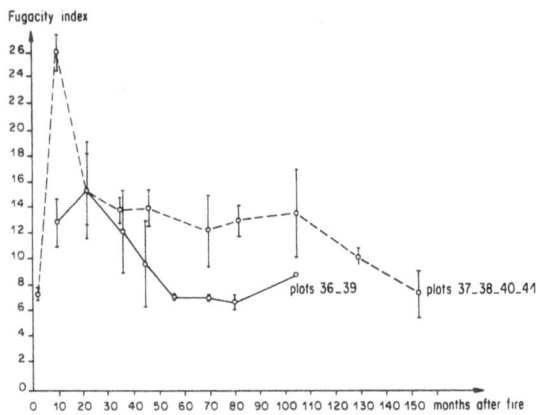

Fig. 14. Development of the fugacity index ($\times$ 100): *Rosmarinus officinalis* garrigues.

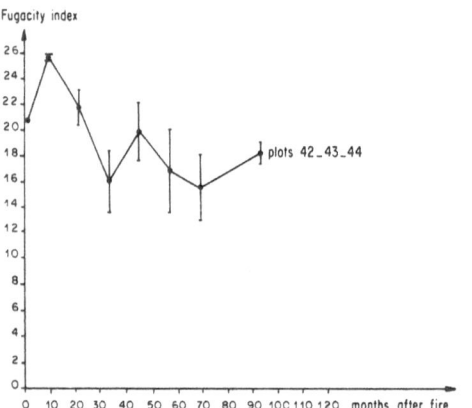

Fig. 15. Development of the fugacity index ($\times$ 100): *Brachypodium ramosum* swards.

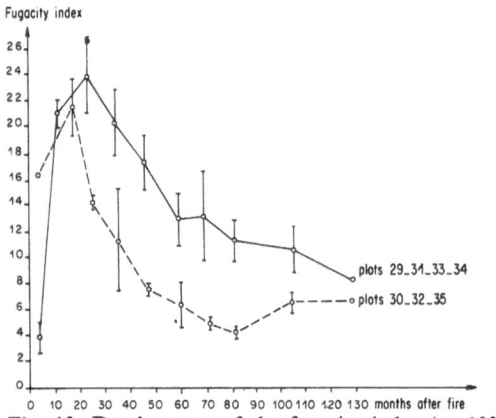

Fig. 13. Development of the fugacity index ($\times$ 100): *Pinus halepensis* woodlands.

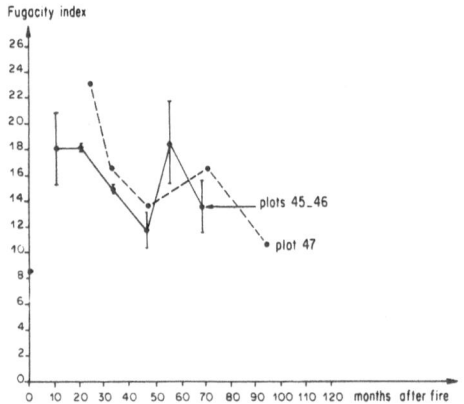

Fig. 16. Development of the fugacity index ($\times$ 100): *Brachypodium phoenicoïdes* swards.

*Spatial variation in floristic composition of communities*

The return of the floristic composition towards a metastable state after fire is a general phenomenon in all communities studied. This state seems very similar to the one which existed before the fire, as appears from the comparison with similar associations already described in literature (Table 1).

A comparison by means of Jaccard's coefficient, between the 'terminal' lists of each of the plots and those of the corresponding associations described by Braun-Blánquet (*Cocciferetum*, 1935, *Quercetum ilicis*, 1937) allows the following conclusions

1) Our *Quercus coccifera* garrigues, are very similar to the *Cocciferetum* of Braun-Blanquet. This result is not surprising, because Braun-Blanquet described this association in the same area and as a post-fire stage.

2) Our *Quercus ilex* coppices are in general rather different from the *Quercetum ilicis*. The explanation may be:

– We studied only the burnt areas, whereas Braun-Blanquet sampled a wide range of stands,

– Recently burnt oak woodlands can be floristically different from older stands. After ten years some stands have evolved again towards a *Quercetum ilicis*.

Would fire lead to identical communities with a common flora on large areas? Apparently the floristic composition of each plot keeps its original character. An indirect way to verify this assertion is to compute the matrix of the Jaccard's similarity coefficients between the observations done in the different plots. The plots characterized by the same dominant species cluster together. Hence we studied the three submatrices determined by the dominance of *Quercus ilex*, *Q. coccifera*, and *Pinus halepensis* + *Rosmarinus officinalis* respectively.

A classification using the simple linkage algorithm was carried out for each of the three groups

To improve the analysis, we computed the compactness and disjunction coefficient (Dunn 1974) for each plot in comparison with the other plots. This coefficient equals the ratio of the lowest similarity value between the lists of plot and the highest similarity between any list of this plot and a list from another plot:

diam $[A]$ = min $S(i, j)$ $i$ and $j \in A$

Sim $[A, B]$ = max. $S(i, j)$    $i \in A$, $j \in B$

$$\forall B \neq A$$

The compactness and disjunction coefficient follows from:

$$CD_A = \frac{\text{diam } A}{\text{Sim } A, B}$$

a value higher than 1 indicates that the group of lists of a plot is compact and well separated from the other groups.

The coefficient was computed in two ways: with, and without, consideration of the first list, when the first observation was done less than one year after a fire.

Most plots appear to form well separated entities. The coefficient of disjunction can reach values up to 2.

Mostly the lower values of the disjunction coefficient were recorded either during the first year or during the second and third year after a fire. In the first case this effect is due to the floristic poorness, and in the second case to the great number of present species. This effect would have been less pronounced if a coefficient would have been used in which species richness would be less important.

In most cases, differences in floristic composition

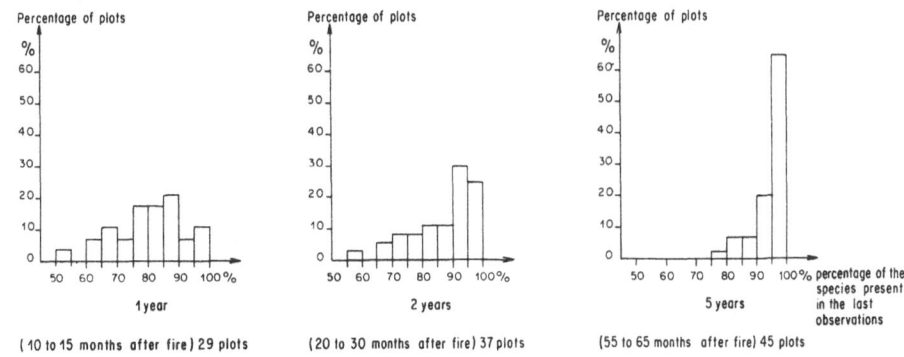

Fig. 17. Relative importance of species present in the terminal communities, measured 1,2 and 5 yr after fire.

55

between plots remain distinctly greater than differences between the lists of one and the same plot. Thus, fire does not appear to modify the floristic diversity of the landscape. At least for the period of observation and the local conditions in Bas-Languedoc, there is no convergence towards a common floristic composition of burnt areas.

### Conclusions

The post-fire development of the floristic composition is very different from that of re-vegetation of bare soil or after abandonment of old fields (Bournerias 1959, Bazzaz 1968, 1975, Mellinger & McNaughton 1975, Guillerm 1978). The opening of the vegetation does allow adventitious species to come in, but these are rapidly eliminated after the return of community species which were present before the fire. During the period of study there is no succession in the sense of substitution of a community by another one, but only a progressive return of previously present species with the temporary superposition of some adventitious species. However, it must be noticed that we have only an inaccurate knowledge of the previous state (only based on burnt snags that remain in place). Among the species which disappeared, some could have played an important role in vegetation dynamics.

Concerning the floristic composition of plots, fire seems to be a rather superficial phenomenon. The return towards a metastable state occurs very quickly and the floristic diversity of the landscape does not seem to be affected.

The *Quercus coccifera* garrigue, a community mainly considered as pyrophytic, is no more stabilized or less diversified than the other communities.

This result can be generalized for entire phytocenoses encountered in Bas-Languedoc, which correspond to different stages of plant successsion. Here, our results are similar to those obtained by American scientists about the Californian chaparral which succeeds itself (Sampson 1944, Horton & Kraebel 1955, Hanes 1971).

The rapid return towards the community existing before fire may be due to the fact that all vegetation of Bas-Languedoc has been influenced by fire for a long time; the maturity levels of communities are low, and species have developed adaptative strategies with regard to fire. Differences existing between communities arise partly from the history of the plots which support them, and

this history is clearly influenced by man's activity, i. e. the use of the land for agriculture, forestry or grazing.

### Summary

Vegetation dynamics after fire was studied in six communities in Bas-Languedoc (Southern France). 47 plots were observed by means of a permanent transect for ten years.

In the first part, we describe floristic richness, species fugacity and the way by which, the 'terminal' community (as defined by the last observation) appears. The dynamics of all these communities follows a simple and general model: floristic richness reaches its maximum during the first two years after a fire, then decreases and becomes stable. Fugacity follows a similar model, whereas the mergence of the 'terminal' community is rapid: one year after fire 70% of the plots have already acquired 75% of the species of the 'terminal' community. There is no succession (in the general sense of the word), but a progressive reappearance of the species belonging to the original community.

In the second part, we study floristic similarities between our plots and corresponding associations as described in literature. It appears that after a fire the floristic diversity of the landscape remains high; while the communities rapidly reach a relative maturity.

In the study area fire seems to be a rather superficial phenomenon; it does not lead to an important modification of the community dynamics, because probably the most frequent species in Bas-Languedoc developed adaptations to withstand fire.

### References

Barry J.P., 1960. Contribution à l'étude de la végétation de la région de Nîmes. Année Biol. 3ème série, 36: 311–550.

Bazzaz F.A., 1968. Succession on abandoned fields in the Shawnee Hills, Southern Illinois. Ecology 49: 924–936.

Bazzaz F.A., 1975. Plant species diversity in old field successional systems in Southern Illinois. Ecology 56: 485–488.

Bharucha F.R., 1932. Etude écologique et phytosociologique de l'association à Brachypodium ramosum et Phlomis lychnitis. Comm. SIGMA. 18, Beihefte Bot. Centralb. 1933, 2: 247–379.

Biswell H.H., 1963. Research in wildland fire ecology in California. Ann. Tall Timbers Fire Ecol. Conf. Proc. 2: 63–97.

Biswell H.H., 1974. Effects of fire on chaparral. In: Kozlowski & Ahlgren (eds.), Fire and ecosystems, p. 321–364 Acad. Press, New York.

Boesch D.F., 1974. Diversity, stability and response to human disturbance in estuarine ecosystems. Proc. 1st. Int. Congress Ecology. The Hague, Sept. 8–14, 1974: 109–114.

Bournerias M., 1959. Le peuplement végétal des espaces nus (essais expérimentaux sur la génèse de divers groupements pionniers). Mem. Bull. Soc. Bot. France, 300 pp.

Braun-Blanquet G., 1936. La lande à Romarin et Bruyère (Rosmarino-Ericion) en Languedoc. Comm. Sigma 48: 8–23.

Braun-Blanquet J., 1936. La forêt d'yeuse languedocienne (Quercion ilicis). Monographie phytosociologique. Mem. Soc. Etud. Sci. Nat. Nîmes, 5, 147 pp.

Braun-Blanquet J., N., Roussine & R., Negre 1952. Les groupements végétaux de la France méditerranéenne. C.N.R.S. Ed. 297 pp.

Daget P. & J. Poissonet. 1971. Une méthode d'analyse phytologique des prairies. Ann. Agron. 22: 5–41.

Dunn J.C. 1974. A fuzzy relative of the Isodata process and its use in detecting compact well-separated clusters. J. Cybernetics 3: 32–57.

Fournier P. 1961. Les quatre flores de France. ed. P. Lechevallier, Paris. 1105 p.

Golley F.B. 1974. Structural and functional properties as they influence ecosystem stability. Proc. 1 st Intern. Congress. Ecology. The Hague, sept. 8–14, 1974: 97–102.

Goodman D. 1974. The validity of the diversity-stability hypothesis. Proc. 1st. Intern Congress Ecology. The Hague, sept. 8–14, 1974: 75–79.

Guillerm, J.L. 1978. Sur les états de transition dans les phytocénoses post-culturales. These Doc. Etat, USTL. Montpellier, 128 pp.

Hanes, T.L. 1971. Succession after fire in the chaparral of Southern California. Ecol. Monogr. 41: 27–52.

Hanes, T.L. & H.W. Jones. 1967. Postfire chaparral succession in Southern California. Ecology: 48: 259–264.

Harper, J.L. 1977. Population biology of plants. Academic Press, New York, 892 pp.

Holling, C.S. 1973. Resilience and stability of ecological systems. Ann. Rev. Ecol. Syst. 4: 1–23.

Horton, J.S. & C.J. Kraebel. 1955. Development of vegetation after fire in the chamise chaparral of Southern California. Ecology. 36: 244–262.

Kornas, J. 1958. Succession régressive de la végétation de garrigue sur calcaires compacts dans la Montagne de la Gardiole près de Montpellier. Acta Soc. Bot. Pol., 27: 563–596.

Kuhnholtz-Lordat G., 1938. La terre incendiée. Essai d'agronomie comparée. ed. La Maison Carrée, Nîmes, 361 pp.

Kuhnholtz-Lordat, G. 1958. L'écran vert. Mem. Mus. Nat. Hist. Natur., série B. Botanique, 9: 276 pp.

Le Houérou, H.N. 1974. Fire and vegetation in the mediterranean basin. Ann. Tall. Timbers Fire Ecol. Conf. Proc. 13: 237–277.

Le Houérou, H.N. 1977. Fire and vegetation in North Africa. Proc. Symp. Environ. Consequences of Fire & Fuel Manag. in Medit. Ecosystems. Palo Alto, Calif., August 1–5, 334–341.

Levy, E. & E. Madden. 1933. The point method of pasture analysis. New. Zeal. J. Agr., 46: 267–279.

Lewontin, R.C. 1969. The meaning of stability. Brookhaven Symp. Biol. 22: 13–24.

Long, G. 1957. La '3 Step method'. Description sommaire et possibilités d'utilisation pour l'observation permanente de la végétation. Bull. Serv. Carte Phytogéogr. B. 2: 35–43.

Long, G. 1958. Description d'une méthode linéaire pour l'étude de l'évolution de la végétation. Bull. Serv. Carte Phytogéogr. B. 3: 107–128.

Margalef, R. 1969. Diversity and stability. A practical proposal for a model of interdependance. Brookhaven Symp. Biol. 22: 25–37.

Mellinger, M.V. & S.J. Mc Naughton. 1975. Structure and function of successional vascular plant communities in Central New York. Ecol. Monographs 45: 161–182.

Naveh, Z. 1974a. The ecology of fire in Israel. Ann. Tall Timbers Fire Ecol. Conf. Proc. 13: 131–170.

Naveh, Z. 1974b. Effects of fire in the mediterranean region. In: Kozlowski & Ahlgren (eds.), Fire and ecosystems, p. 321–364 Acad. Press, New York.

Naveh, Z. 1975. The evolutionary significance of fire in the mediterranean region. Vegetatio, 29: 199–208.

Naveh, Z. 1977. The role of fire in the mediterranean landscape of Israel. Proc. Symp. Environ. Consequences Fire, Fuel Manage in Medit. Ecosystems, Palo Alto, Calif. August 1–5, 299–306.

Orians, G.H. 1974. Diversity, stability and maturity in natural ecosystems. Proc. 1st Intern. Congress Ecol., The Hague, Sept. 8–14, 1974: 64–65.

Patric, J.H. & T.L. Hanes. 1964. Chaparral succession in a San Gabriel Mountain area of California. Ecology, 45: 353–360.

Pavillard, J. 1935. Eléments de sociologie végétale. Act. Scient. Indust., 251: 102 p. Herman et Cie, Paris.

Sampson, A.W. 1944. Plant succession on burned chaparral lands in northern California. Bull. no. 685. California Agr. Exp. Sta., Berkeley, 144 pp.

Soroceanu, E. 1936. Recherches phytosociologiques sur les pelouses mésoxérophiles de la plaine langue docienne (Brachypodietum phoenicoïdis). Comm. SIGMA 40, 250 pp.

Stone, E.C. & G. Juhren. 1951. The effect of fire on the germination of the seed of Rhus ovata. Amer. J. Bot. 38: 368–372.

Sweeney, J.R. 1956. Responses of vegetation to fire. A study of the herbaceous vegetation following chaparral fires. Bot. Publ. Univ. Calif. 28: no. 4: 143–250.

Sweeney, J.R. 1967. Ecology of some 'fire type' vegetation in northern California. California Tall Timbers Fire Ecol. Conf. Proc. 7: 110–125.

Vogl, R.J. & P.K. 1972. Fire and manzanita chaparral in the San Jacinto mountains, California. Ecology 53: 1179–1188.

Accepted 20 December 1979

# DEVELOPMENT OF SPECIES DIVERSITY IN SOME MEDITERRANEAN PLANT COMMUNITIES*

C. HOUSSARD, J. ESCARRÉ & F. ROMANE**

Centre d'études phytosociologiques et écologiques Louis Emberger (C.N.R.S.) B.P. 5051, 34033 Montpellier Cedex, France

Keywords:
Diversity, Forest, French mediterranean region, Old-fields, Phenology, Solar radiation, Stratification, Structure, Succession

## Introduction

Changes in plant species composition during succession are interpreted as the result of a dynamic equilibrium between colonization, persistence and extinction of species.

Early successional communities are composed of species with a short life span (Odum 1969, Salisbury 1942). In more mature communities, most of the species are chamaephytes and phanerophytes which tend to persist together for a long time. Earlier dynamic models of ecosystems as suggested by Margalef (1957, 1968) and Odum (1969) mainly postulate that, during succession, the biomass increases, as primary production does, while diversity originally increases and decreases toward the final stage of succession afterwards; fluctuations are damped and stability increases. These earlier hypotheses have been criticized by Drury & Nisbet (1973), May (1973), Colinvaux (1973) and Horn (1974). In recent studies (Connell & Slatyer 1977, Grime 1977, Grubb 1977, Whittaker 1977, Whittaker & Levin 1977, Huston 1979) one has tried to synthesize empirical results and proposed more elaborate models to explain successional patterns and community diversity, stability and organization.

*Nomenclature follows: P. Fournier (1961). Les quatre flores de France for the Gramineae and Liliaceae, Flora Europaea for the other taxa.
** The support of the Délégation de la Recherche Scientifique et Technique through the grants N567, N633, N674 (Structure et dynamique des formations à Chêne pubescent en zone bioclimatique méditerranéenne) is acknowledged. We thank J. Blondel, R. Bonhomme and M. Ducrey for technical and scientific assistance. We are greatly indebted to Dr. E. van der Maarel for his helpful improvements of this text.

Two ecological problems were especially emphasized:
1. The substitution of species during the succession.
2. The cohabitation of species in mature communities.

Most studies on plant succession have been carried out in non-mediterranean countries. In this paper our main purpose is to compare the results achieved in the field of vegetation dynamics with those in the French Mediterranean region and to treat in particular:
1. The evolution of species richness in the course of an old field succession, ending in an oak forest (*Quercus pubescens*) and the mechanisms causing a substitution of species. Here our investigations concentrate on a. lifeforms as an expression of stature and species life span, and b. dispersal means as an expression of the capability of species to colonize a new habitat.
2. The tendency of species to evolve towards coexistence in more mature communities such as a *Quercus pubescens* forest. This has been investigated by means of three variables: a. the stratification of species and foliage, b. the relation of species to a resource gradient such as light, c. the phenology of species. These variables may be considered as indicators of community organization and resource allocation.

Our contribution consists of two parts, dealing with 1. changes during succession towards a *Quercus pubescens* forest, 2. community structure in relation to the organization of two *Quercus pubescens* forests representing the final stage of succession.

## Study area

The investigation sites are located in the region of the

Hautes Garrigues du Montpelliérais, North of Montpellier, at an elevation of 230 m. The average annual precipitation amounts to 120 cm with maxima in spring and autumn. Evergreen forests dominated by *Quercus ilex*, deciduous forests dominated by *Quercus pubescens*, and mixed formations including both species are considered as the mature formations of the region. This mature vegetation was phytosociologically described as a transition between the *Quercetum ilicis galloprovincialis* and the *Querceto – Buxetum* (Braun-Blanquet 1936, Braun-Blanquet et al. 1952). In coppices of *Quercus pubescens* we find some septentrional species which extend southward of their usual habitat because of the special ecological conditions of the deciduous coppice (Blondel 1941).

The succession study was carried out in adjacent abandoned vineyards and olive groves on marly soil, in two localities at 30 km distance in Le Causse de la Selle and in La Boissière, with a *Quercus pubescens* forest as the potential vegetation. Data were collected during 1977 and 1978 from sites on 19 old fields which were abandoned for 4, 8, 9, 10, 12, 14, 15, 16, 17, 20, 21, 25, 30, 50, 54, 58, 62, 72, resp. 125 years. Some fields were reinvestigated in the second year. At La Boissière only the earlier stages from 1 to 25 years were analysed.

The two old field series are very similar in major successional trends and constitute parallel series according to factorial analysis of correspondances (Benzecri et al. 1973) based on the floristic composition of the sites (see appendix).

Horizontal, vertical and temporal patterns of plant species in the final stage, i.e. the forest, were studied in two *Quercus pubescens* stands, sampled in the La Rouvière forest at 50 m distance from the La Boissière old fields. The two stands had no human influence after cutting in 1945. The forest was earlier destroyed by fire in 1918 and 1923; while grazing occurred in the area during the 17th century and between 1840 and 1930 (Vidal 1979).

**Methods**

Two different sampling schemes were used for the old field sere and the forest. In the old field sere, species richness of all sites was analyzed in areas of (20 m × 1 m) for herbaceous species and (20 m × 10 m) for woody species. Cover of species was taken as importance value and determined by the line interception method (Canfield 1941, Buell & Cantlon 1950) along a 20 m transect. The age of the fields (number of years after abandonment) was determined according to information from the landowners and by means of woody species if present on the site.

To identify species dispersal means we used the classification of diaspores (Table 1) proposed by Molinier & Müller (1938) and van der Pijl (1972). Only three dispersal types were analyzed: barochory, anemochory and zoochory.

Life forms types sensu Raunkiaer were determined through own observations and further based on reviews of species life forms in the mediterranean region by Blondel (1941) and Braun-Blanquet et al. (1952). The frequency of dispersal and life form types are expressed as the number of species belonging to a type relative to the total number of species in the sample.

In the forest communities vertical and horizontal structure was studied along a 32 m transect. The presence of each species was registrated in units of 10 cm height and 10 cm length from the zero to the 2 m level, along the transect.

The vertical vegetation profile above 2 m was determined with the optical point-quadrat method developed by McArthur & Horn (1969), using a 50 mm camera

Table 1. Classification of dispersal types according to Molinier & Müller (1938) and van der Pijl (1972) slightly simplified.

| | |
|---|---|
| BAROCHORES | Diaspores fall by their own weight |
| ANEMOCHORES | Diaspores moved by wind |
| | with – LIGHT GLIDERS |
| | – HEAVY GLIDERS |
| | – ROLLERS |
| | – PROJECTORS |
| ZOOCHORES | Diaspores normally dispersed by animals |
| | with – EPIZOOCHORES: diaspores provided with mechanisms permitting the adherence to animals |
| | – ENDOZOOCHORES: diaspores ingested by animals |
| | – with ELAIOSOMES: diaspores with aril or caruncle |
| | – DYSZOOCHORES: diaspores normally digested, accidentally transported |
| AUTOCHORES | Diaspores moved by their own weight |
| | with – WALKERS |
| | – PHYSIOLOGICAL PROJECTORS |
| | – MECHANICALS PROJECTORS |
| | – CRAWLERS |
| HYDROCHORES | Diaspores moved by water |
| | with – NAUTOHYDROCHORES |
| | – OMBROHYDROCHORES |

described by Blondel & Cuvillier (1977) with a telephoto lens having a point marked on the center of the viewfinder screen.

The height above the lowest leaf covering the point is determined by focusing the lens on that leaf and by reading the distance off the lens mount. The distribution of the height measures to the lowest leaf can be transformed into estimates of vertical distribution and total amount of foliage (McArthur & Horn 1969). Aber (1979) concluded that this method does not provide accurate estimates of the leaf area index, but still it is a rapid and accurate way to estimate canopy foliage-height profiles in broad-leaved forests.

Daily extinction of global radiation for the summer solstice in the understory of the two stands was estimated with a mathematical model described by Ducrey (1975). The principle of the method is 1. to estimate the potential radiation at the measurement point with semi-empirical formulas proposed by Krochmann (1973) and Dogniaux & Lemoine (1976), 2. to determine the distribution and gap frequency seen from the measurement point on hemispherical photographs, and 3. to estimate the daily fractional penetration of the global radiation reaching this point by taking into account the radiation coming directly through the gaps of the canopy (Ducrey 1975) and the potential radiation at the point of measurement.

Hemispherical photographs were made in 8 horizontal positions along the 32 m long transect at a 2 m level in La Rouvière – 1 above the shrub layer, and at a 50 cm level in La Rouvière – 2 above the herbaceous layer.

The main phenophases of the dominant species in the understory (flowering stage, fruiting stage, fruit ripening stage and fruit disappearance) were observed as to their duration.

### Results

*General description of the plant communities*

The development sequence of the old fields sere is as follows: during the first year after abandonment the fields are dominated by spring annuals such as: *Stellaria media*, *Cerastium pumilum*, *Cerastium glomeratum* and *Cardamine hirsuta*; by *Avena barbata* in June and July; by *Conyza canadensis* in September. In the first 15 years perennial species are progressively taking over.

Patches of one dominating perennial species, surrounded by annual species from earlier stages, can be observed.

Dominant species are: *Calamintha nepeta*, *Thymus vulgaris*, *Lavandula latifolia*, *Rubus ulmifolius*, *Brachypodium phoenicoides*. The latter species progressively invades declining patches of *Thymus vulgaris* and *Lavandula latifolia*.

At this stage we can observe the spread of individuals of shrubs: *Rosa canina*, *Crataegus monogyna*, *Phillyrea angustifolia*, *Lonicera etrusca*, *Sorbus domestica*, *Acer monspessulanum*, *Ruscus aculeatus*, *Prunus spinosa*, and (in La Boissière), *Cornus sanguinea*.

In 15- to 50-years old fields the grass *Brachypodium phoenicoides* dominates, forming thick clumps which exclude the other species, notably the *Labiatae*.

In abandoned olive groves, the herbaceous vegetation is dominated by *Labiatae* (*Calamintha nepeta*, *Satureja montana*, *Thymus vulgaris* and *Lavandula latifolia*, later succeeded by *Brachypodium phoenicoides*. Near most of the old olive trees we find one or a few *Juniperus oxycedrus* individuals. Interestingly, at their basis, typical forest species: *Ruscus aculeatus*, *Clinopodium vulgare*, *Tamus communis*, *Lonicera etrusca*, *Clematis vitalba*, *Hedera helix*, *Rubus ulmifolius* germinate and develop.

In 50- to 125-year old fields, *Quercus pubescens* is becoming the dominant species. Shrub such as *Prunus spinosa*, *Pistacia terebinthus*, *Buxus sempervirens* become more abundant in the understory. Heliophilous species from previous stages disappear.

The two forest communities appear to have different understories. The first one, La Rouvière-1, exhibits an understory with a high number of woody species, with *Cornus sanguinea*, *Prunus spinosa* and *Phillyrea angustifolia* dominating below the height of 2 m, *Crataegus monogyna* and *Cornus sanguinea* above 2 m.

The age of *Phillyrea angustifolia* was determined as 40 yr, *Cornus sanguinea* and *Prunus spinosa* 14–18 yr, *Crataegus monogyna* 22, resp. 32 yr.

The canopy height is 9 m and the stem density 4200 trees/ha with a productivity of 4.6 m$^3$/ha/yr (Fernandez 1978). The favourable soil conditions (deep and well drained) may explain the richness of the woody species in the understory.

In the second site, La Rouvière-2, the understory is dominated by herbaceous species: *Festuca spadicea*, *Brachypodium pinnatum*, *Bromus erectus*, *Carex glauca*. Some patches with *Phillyrea angustifolia*, *Lonicera etrusca* and *Clematis flammula* appear locally. Here the canopy height is 7 m, the stem density 3400 trees/ha with a productivity of 2.6 m$^3$/ha/yr (Fernandez 1978). The herbaceous character of the understory may be explained by the

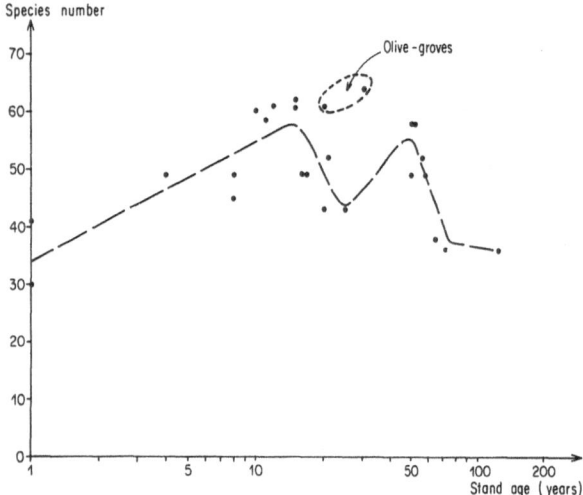

Fig. 1. Changes in species richness with successional age in the Montpellierais region. Points encircled by a dotted line correspond to two abandoned olive groves which have not been taken into account for the drawing of the dotted curve (see text).

unfavourable soil conditions, i.e. a stony soil which limits the development of trees.

## Results from the old field sere

### *Development of plant species richness and dominance*

Species richness fluctuates in time (Fig. 1); 1–15 year increase, between 20 and 25 year decrease, around the 50th year increase, finally in the remaining years, from 50–125 years decrease. Diversity (measured as $N_1 = 2^H$, $H$ being Shannon's index, cf. Hill (1973a) and equitability ($N_1/N_0$, $N_0$ being total number of species) follow the same trends (Escarré 1979). Two fields, abandoned 20- and 30-year old olive groves, exhibit a high species richness, whereas in other samples, trends show a decrease in diversity at this stage.

The relation between species richness and species dominance is expressed in the form of dominance-diversity curves (Whittaker 1965). In the case of the old field sere log relative cover values are plotted against abundance rank of the species. Relative cover is given as the percentage contribution of species to the total cover of all species determined with the line-intercept method. The resulting species graphs, arranged according to community age are given in Fig. 2.

The relation for the 1-year old community, approxi-

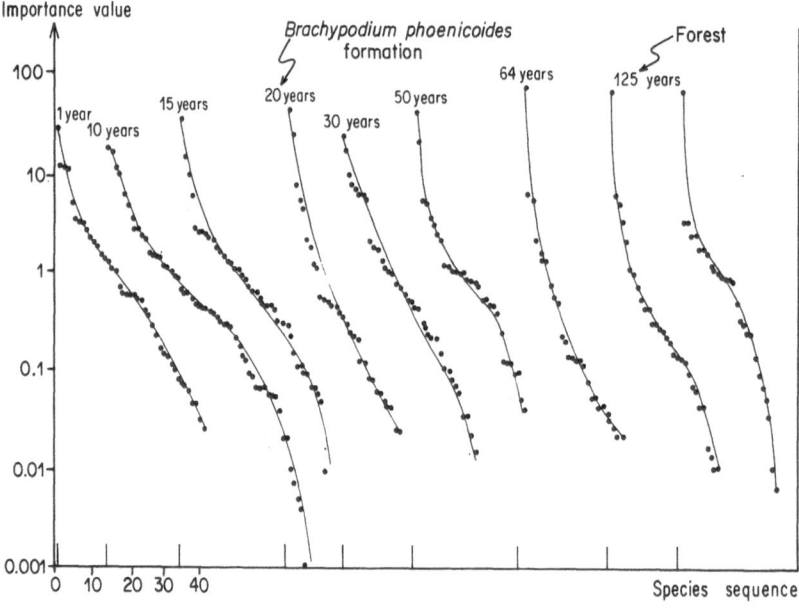

Fig. 2. Dominance-diversity curves for successional communities in Le Causse de la Selle. The points represent the species, plotted as relative cover importance against the species rank in the sequence of species from most to least dominant. To make the figure clearer, the curves have been arbitrarily spaced out. Positions of their origins on the abscissa are indicated by vertical lines along the low border of the figure.

mates a geometric series; for 9–15 year old fields it tends toward a log-normal distribution. The curves for the 20–30 year old communities become steeper and approach again a geometric series, but in the 50-year old field, the curve suggests again a log-normal distribution. The curve representing the 64-year old field has a steep slope and approximates a geometric series while for successively older communities (up to 125-year old ones) the curves approximate a log-normal distribution.

The geometric curves show strong dominance of some species and confirm the hypothesis of niche-pre-emption (Motomura 1932, Whittaker 1965, 1969, 1972). One dominant species may occupy a similar fraction of the total niche space, the second species may occupy a similar fraction of the remaining space, and so on. Such a form is often exhibited by vascular plant communities with a low diversity (Whittaker 1972). A log-normal distribution of importance values suggests that species populations are determined by a number of independent variables (May

1975). In such cases few important species, few very rare species and a lot of species of intermediate value will occur. Such a distribution is usually found in vascular plant communities with a high diversity (Whittaker 1965).

In our data, each diversity maximum is linked with a log-normal distribution of importance values. An increase in species richness corresponds to a decrease in dominance, as also observed by Whittaker (1965) and Bazzaz (1975).

*Changes in species turn-over rate*

We have estimated the percentage of species occurring at a particular stage that is still present in later stages (Fig. 3). Shugart & Hett (1973) proposed a formula to express the substitution rate of species in various successions. This formula has been applied by Escarré (1979) to the same group of old field sere; it leads to the same interpretation as our more simple approach. The species turn-over rate is very high during the first 50–60 years of the sere. The

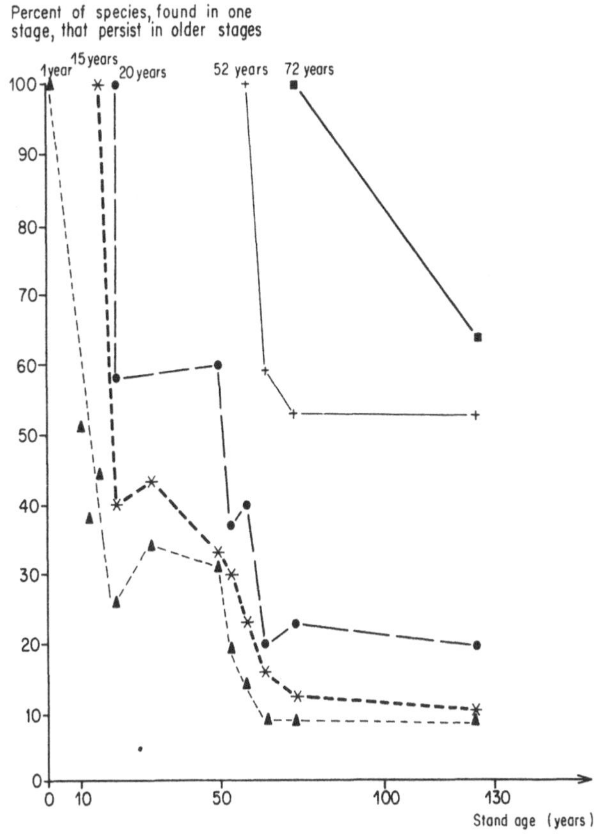

Fig. 3. Species turnover in the course of succession, at Le Causse de la Selle (Hérault) for five stages (1, 15, 20, 58 and 72 years old) belonging to the same successional sequence.

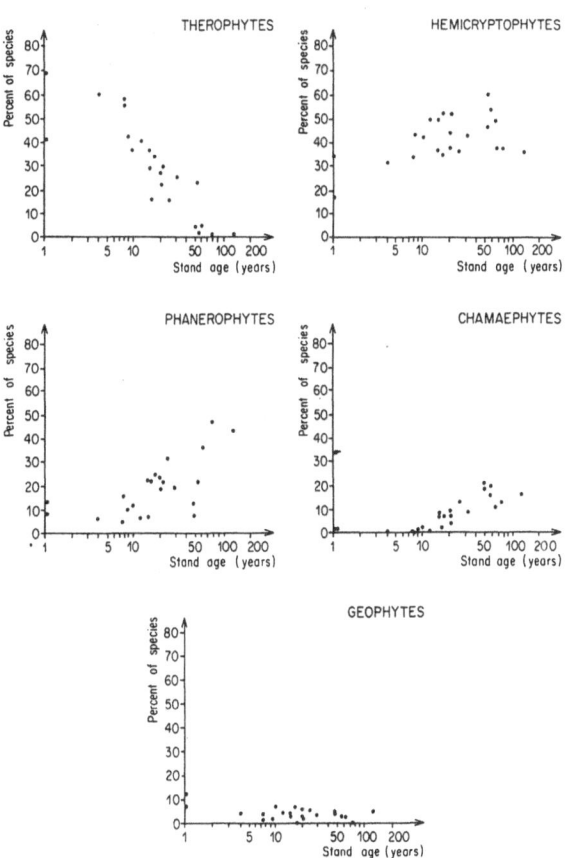

Fig. 4. Percentage of species belonging to each life-form on successional old fields of different ages.

older stages are more similar with regard to their species composition. In the sense of species constancy the ecosystem can be considered more stable (Margalef 1957). Similar results were obtained by Margalef (1957), Odum (1969) and Shugart & Hett (1973).

*Changes in the life form spectrum*

Five life form types were distinguished: hemicryptophytes, geophytes, therophytes, phanerophytes and chamaephytes. Their development in the course of the succession is given in Fig. 4. Therophytes and phanerophytes show the clearest trends; therophytes are abundant in the first stages from 1- 15-year old. In later stages they decrease. The phanerophytes behave oppositely. Nevertheless, they are found already in the first stages. Chamaephytes are not much represented at the 2- to 10-year old stage and they regularly increase beyond this period. The trend followed by hemicryptophytes is not very clear; they represent ca. 30 % of the total number of species during all the stages. We notice that in 15-year old communities all life forms are represented, which is not the case for the earlier or older communities.

*Changes in the spectrum of dispersal types*

The trends of the three major dispersal types are clear (Fig. 5): zoochores and barochores increase regularly in the course of succession. Anemochory is very well rep-

resented in 1 to 15-year old fields and then decreases. According to van der Pijl (1972), the barochores form a group including species with diaspores having no evident dispersal means. We may remark that barochoric species are often also dispersed by animals, notably ants and rodents. The complementary relation between these two dispersal types is important; it contributes to the maintenance of species in the ecosystem and to the colonization of the species in other communities. The three dispersal types are all present in the 15-year old fields, which is not the case for earlier or older communities.

**Results for the forest communities**

In the forest stage of succession, the species turn-over rate is low whereas perennial growth-forms (phanerophytes and hemicryptophytes) are dominant. Species tend to persist for a long time. In order to understand the way species coexist in plant communities, we have examined the two *Quercus pubescens* forests La Rouvière-1 and La Rouvière-2 as follows. We described the way various species divide up space and we investigated if understory diversity could be correlated with the forming of patterns of species, having a similar resource gradient exploitation (light) or synchronized phenological behaviour.

*Below the 2 m level*

If we consider the total number of species present in each 10-cm height interval, the profile (Fig. 6b) suggests only one evident stratum of vegetation in the first 20 cm above the soil level; in La Rouvière-1, it affects 65 % of the total species frequencies, in La Rouvière-2 even 85 %.

When we consider species modes, variance/mean ratios are not different from 1 (Table 2), indicating a random distribution of species in the understory, except in La Rouvière-2-with 3 units – where the $V/m$ ratio is lower than 1 indicating a hostile or competitive distribution of species. In the case where centres of gravity are considered, the degree of aggregation is always greater than 1 and higher with 18 units than with 3 units, in the two stands, indicating an aggregative distribution of the centres of gravity of species. However, the degree of aggregation is never very high. We may thus conclude that species have a tendency to concentrate their distribution of foliage (as expressed by the centre of gravity) in only a few layers. This is not so, however, if we consider the mode of this distribution as observed in the foliage profiles (Fig. 6b).

In La Rouvière-1, six species concentrate their centres of gravity between 20 and 50 cm: *Crataegus monogyna*

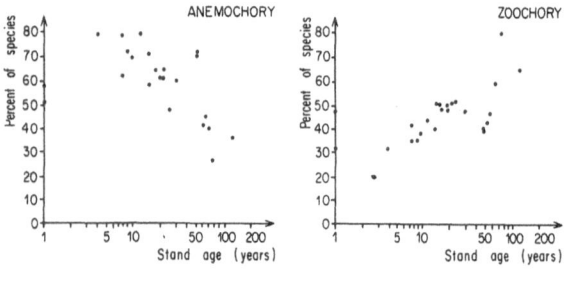

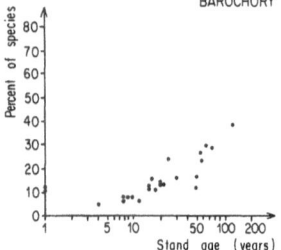

Fig. 5. Percentage of species belonging to each dispersal class on successional old fields of different ages.

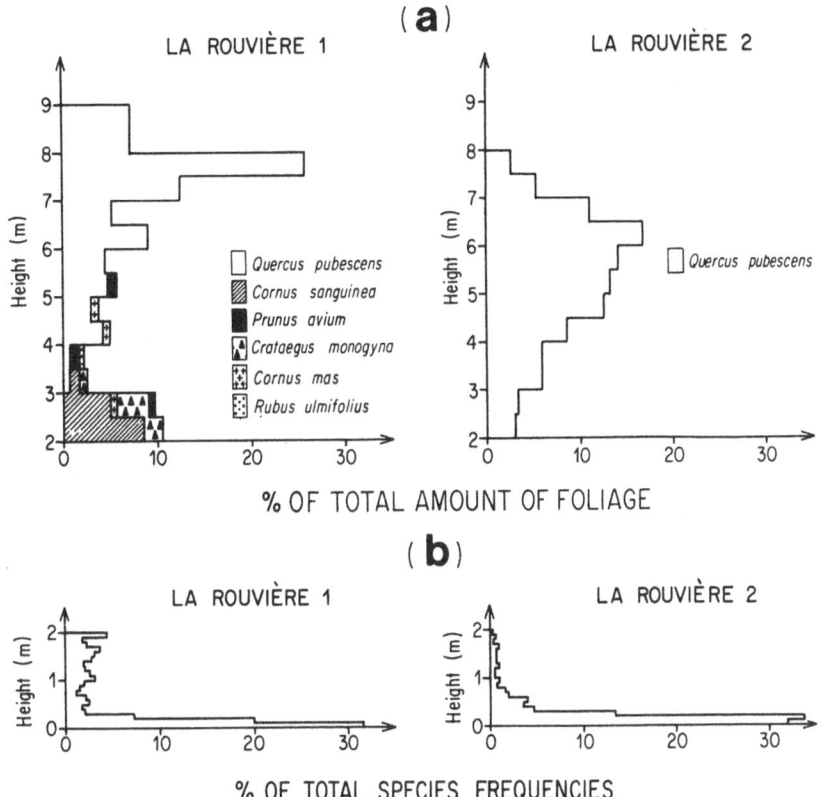

Fig. 6. Foliage height profiles for two *Quercus pubescens* stands expressed as the percentage of the total foliage density every 50 cm interval above 2 m (6a) and as the percentage of the total species frequencies every 10 cm interval between 0 and 2 m (6b).

(seedling), *Rosa sempervirens*, *Prunus spinosa*, *Cornus sanguinea* (seedling), *Rubia peregrina* and *Phillyrea angustifolia*. Five species concentrate their centres of gravity between 1.40 and 1.50 m: *Crataegus monogyna*, *Prunus*

Table 2. Variance: mean ratio for the distribution of modes and of the centres of gravity of the species frequencies into vertical units between a 20 cm and a 2 m level in the two *Quercus pubescens* stands of La Rouvière. Two sections of the vertical axis have been taken into account: 18 units (10 cm high) and 3 units (20-50, 50-100, 100-200 cm).

| 18 vertical units | LA ROUVIERE 1 | LA ROUVIERE 2 |
|---|---|---|
| V/m for the modes | 1.5 | 0.7 |
| V/m for the centres of gravity | 1.8 | 2.2 |
| 3 vertical units | | |
| V/m for the modes | 1.5 | 0.3 |
| V/m for the centres of gravity | 1.8 | 1.5 |

*spinosa*, *Amelanchier ovalis*, *Cornus sanguinea*, *Phillyrea angustifolia*. These two concentrations of centres of gravity can be considered as two representing strata.

In La Rouvière-2, seven species concentrate their centre of gravity between 20 and 50 cm: *Peucedanum cervaria*, *Phillyrea angustifolia*, *Asphodelus cerasifer*, *Festuca spadicea*, *Lonicera etrusca*, *Clematis flammula*, *Arrhenatherum elatius*. Only *Festuca spadicea* is really abundant in this layer. Between 0.50 and 1 m, four species have their centres of gravity: *Quercus ilex*, *Pteridium aquilinum*, *Quercus pubescens* and *Phillyrea angustifolia*. These four species are not very abundant in this layer.

*Above the 2 m level*

In La Rouvière-1, 50 % of the total amount of foliage is located above 7 m and consists of *Quercus pubescens* (Fig. 6a). *Cornus sanguinea* and *Crataegus monogyna* contribute to 20 % of the total amount of foliage between 2 and 3 m. However, the modes of the total amount of foliage of

65

each species do not coincide with each other in the same unit (between 2 and 2.5 m for *Cornus sanguinea*, between 2.5 and 3 m for *Crataegus monogyna*). In La Rouvière-2, only *Quercus pubescens* is present; its mode of distribution of the foliage is concentrated between 4.5 and 7 m. La Rouvière-1 stand exhibits a more complex vertical structure – with species layers – than La Rouvière-2 stand, which has a relatively simple vertical structure with only one species dominating in the layers.

## Horizontal pattern of species

The horizontal pattern of species has been brought in relation with the gradient of light in the understory. The La Rouvière-1 understory has a high woody species diversity below 2 m. Consequently we have examined the pattern of abundance of the dominant shrubs along the

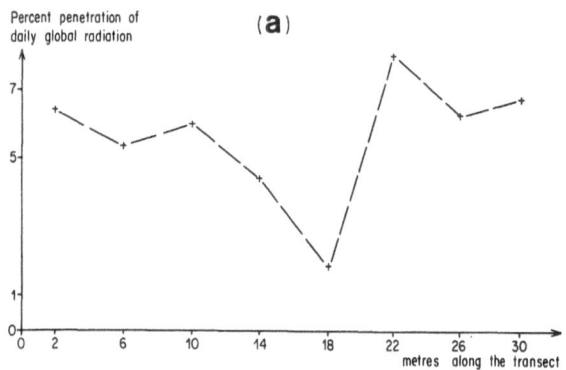

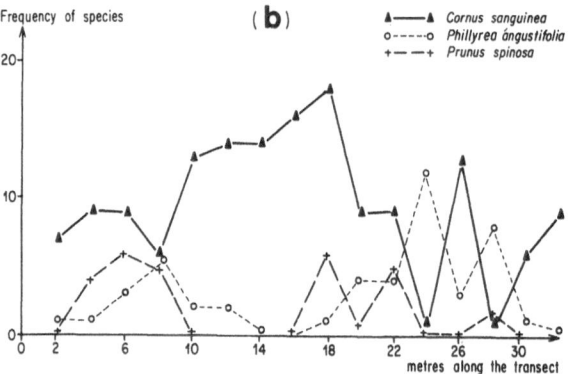

Fig. 7. Relations between the frequency of the dominant shrub species (7b) and the daily fractional penetration of global radiation estimated with a model (7a), at a 2 m level, from an overcast sky, on June 21, in La Rouvière 1, along the 32 m long transect.

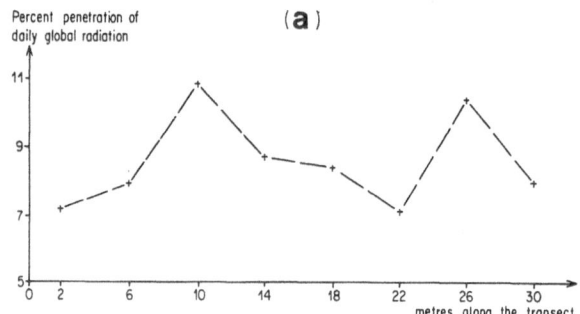

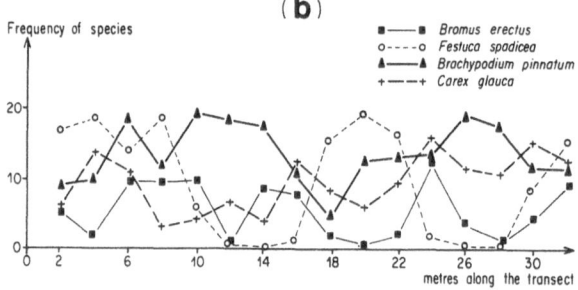

Fig. 8. Relations between the frequency of the dominant herbaceous species (8b) and the daily fractional penetration of global radiation estimated with a model (8a), at a 50 cm level, from an overcast sky, on June 21, in La Rouvière 2, along the 32 long transect.

transect in relation to the global radiation, entering the 2 m level on June 21. *Cornus sanguinea* and *Phillyrea angustifolia* exclude each other (Fig. 7b). The former is a deciduous species, the latter an evergreen. In the middle of the transect, where extinction of light is maximal (Fig. 7a), *Cornus sanguinea* is dominant and excludes other species. At the two extremities of the transect, where extinction of light is low, dominance of *Cornus sanguinea* is reduced, and an increase in species richness can be observed (Fig. 7).

As the La Rouvière-2 understory is rich in herbaceous species, we also examined the variations of the abundances of the four dominant grasses, in relation to global radiation above this herbaceous stratum on June 21. Mosaics of *Festuca spadicea* and *Brachypodium pinnatum* that seem to exclude each other give evidence of the specific structure of the understory (Fig. 8b). Abundance of *Brachypodium pinnatum* is maximal when light extinction is low (Fig. 8). The opposite is the case with *Festuca spadicea*. Patterns of *Bromus erectus* and *Carex glauca* seem to be imposed upon by the two preceding species (Fig. 8b). Consequently the light climate in the understory may account for the niche differentiation of the dominant shrubs and herbaceous species in the two stands and give evidence of diversity.

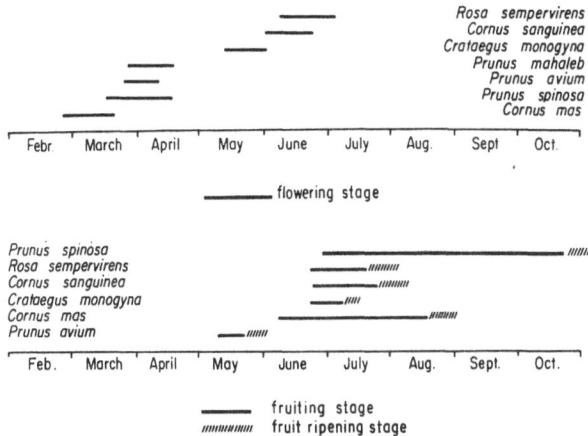

Febr.  March  April  May  June  July  Aug.  Sept  Oct.

——— flowering stage

——— fruiting stage
////// fruit ripening stage

Fig. 9. Major phenophases in 1978 for the dominant species in the understory of La Rouvière 1.

## Phenological patterns of species

The phenological patterns of the species in the understory have been studied by recording the major phenophases in the shrub strata of La Rouvière-1 and in the herbaceous strata of La Rouvière-2. In La Rouvière-1 an asynchronism in flowering initiation and in fruit ripening stage has been noticed (Fig. 9). Four flowering periods may be distinguished, each of them being characterized by the following species: 1. *Cornus mas*. 2. *Prunus spinosa*, *Prunus avium*, *Prunus mahaleb*. 3. *Crataegus monogyna*. 4. *Rosa sempervirens*, *Cornus sanguinea*.

Five periods may be determined for the fruit ripening stage (Fig. 9): 1. *Prunus avium*. 2. *Crataegus monogyna*. 3. *Cornus sanguinea*, *Rosa sempervirens*. 4. *Cornus mas*. 5. *Prunus spinosa*.

An optimum phase with most of the species fruiting

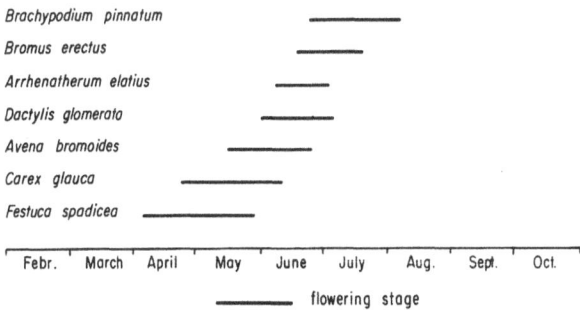

Febr.  March  April  May  June  July  Aug.  Sept.  Oct.

——— flowering stage

Fig. 10. Flowering stages in 1978 for the dominant herbaceous species in the understory of La Rouvière 2.

without ripening yet begins during the 3th and 4th week of June. It corresponds to the period when daily radiation is longest and when *Quercus pubescens* foliation has not finished. In La Rouvière-2 we observed an asynchronism in the flowering initiation of grass species, which extends from the end of April to the end of June. (Fig. 10). Two main groups with flowering synchronism concerning the dominant species, may be considered: *Festuca spadicea* and *Carex glauca* for the first one, and *Brachypodium pinnatum* and *Bromus erectus* for the second one.

We have thus observed a spreading of the major phenophases of species in the two sampled forest stands.

## Discussion and conclusions

The fluctuation of species richness in the course of the studied succession may be due to the presence of social species which appear in successive waves, depending on their dispersal power and their presence around the field. In earlier stages of the succession, species are essentially dispersed by wind (Fig. 5). As wind is an efficient dispersal agent, but a random one, these species may produce a great number of seeds (species with 'r-selection') to ensure germination (McArthur & Wilson 1967). Lewontin (1965) showed that a high 'r' can be achieved by reducing the length of the development rather than by increasing fertility. In particular, pioneer species are essentially annual (Fig. 4).

In stages from 1 to 15-year old the increase in species richness (Fig. 1) is related to a rapid turn-over rate of species (Fig. 3). Hence there is a progressive partitioning of individuals among more and more species. All the life forms and dispersal means are present and a mixture of both early and late successional species can be observed. Considering the case of an old-field sere in Central New York, Mellinger & McNaughton (1975) infer that a high level of species diversity would be brought about by a local differentiation in soil properties around individual plants (nutrient depletion or enrichment, accumulation of allelopathic chemicals by *Labiatae* for example); the production of such resource inhibitor fields in soil would increase the microgeographic heterogeneity, providing loci for the successful penetration of communities by the individuals of a new species.

Between 15 to 25 years of succession the decrease in species richness (Fig. 1) develops with a maximum dominance by *Brachypodium phoenicoides*. The same coincidence of the decrease of species richness with the dominance of one herbaceous species was noticed by Bazzaz (1975) in

67

Southern Illinois in the case of *Andropogon virginicus* from the 4th to the 10th year of the succession and by Nichols & Monk (1974) in Piedmont (Georgia) in the case of *Andropogon scoparius* from the 10th to the 20th year.

Between 25 and 50 year of succession the progression toward a log-normal curve (Fig. 2) and the increase in species richness (Fig. 1) coincide with the wilting of the foot of *Brachypodium phoenicoides* tussocks. In the same time *Quercus pubescens* individuals settled on dry stone walls, are a source of diaspores for abandoned fields. These individuals cast shadow and cause some species to etiolate such as *Brachypodium phoenicoides* and *Bromus erectus*. Thus, this process gives new species a possibility to colonize the cleared space.

In 50- to 125-year old fields, *Quercus pubescens* progressively tends to be the dominant species; a decrease in species richness with a stabilization at a level of 40 species per releve can be observed. The decrease in species richness with tree settlement is due to the suppression of heliphilous species with the closing of the canopy. The reduction in species diversity with the appearance of woody species was noticed by Whittaker (1972). Woody species of higher strata are barochorically dispersed. For *Quercus pubescens* this is the only efficient mechanism of colonization. Here the problem rises of the vulnerability for attacks of some predatory animals, among which rodents and herbivorous insects, which destroy seeds and seedlings. The fall period of fruits is usually short and Janzen (1970) expresses the hypothesis that species have adopted a 'repletion' strategy: predatory animals cannot completely destroy the great number of seeds which have been produced within a very short lapse. In the temperature zone, many authors (Salisbury 1942, Sharp & Sprague 1967, Murphy 1968) noticed a great fluctuation of seed production. This fluctuation is synchronized between all individuals belonging to the same species; a year of seed overproduction secures a stock of seeds sufficient for reproduction.

In 125-year old fields the dominance-diversity curves tend toward a log-normal form (Fig. 2). This may be due to the development of shrub species in the understory.

The exceptionally high species richness in 20–30 year-old abandoned olive groves (Fig. 1) may be related to tree and herb layers that enhance vertical heterogeneity. Tree trunks are used as perching places by birds. This favours the settlement of endozoochorically dispersed species near tree trunks. Old vine trunks produce the same effect and we frequently observed woody species, with endozoochoric dispersal near the trunk. Under the persistent species, such as *Juniperus oxycedrus* we found microhabitats, distinct from the rest of the field, where a humus layer accumulates and soil maturation rapidly proceeds. Thus conditions are very similar to those of the forest understory and we noticed, under persistent species, the presence of species belonging to more mature stages of the succession (*Tamus communis*, *Hedera helix*, *Ruscus aculeatus*, *Lonicera etrusca*). This process was observed by Bazzaz (1968) and is very similar to the nucleation process described by Yarranton & Morrison (1974) for *Juniperus communis* ssp. *virginiana* which acts as a nucleus for the micro-growth patches of persistent species. This is a good example of a high species diversity due to heterogeneity, since heterogeneity of environment allow satisfaction of the requirements of many species within a community (Whittaker & Levin 1977).

Our results on diversity trends in the course of succession seem to be similar to those of Auclair & Goff (1971), Reiners et al. (1971), Holt & Woodwell, unpublished, in Whittaker (1975), in the sense of an absence of a monotonous increase in diversity after forest establishment. They differ from those of Nicholson & Monk (1974), Bazzaz (1975) and Mellinger & McNaughton (1975). These authors found a regular increase in species diversity during succession. They interpret this trend as an indication of the compression of niches realized through competition (Mc Arthur & Wilson 1967) and species packing (Mc Arthur 1969, May & Mc Arthur 1972). This hypothesis does not help to understand our pattern of diversity.

An alternative hypothesis, formulated by Huston (1979) explains our diversity trends better. This author infers that most communities exhibiting a high species diversity are kept in a non-equilibrium state where a competitive equilibrium is prevented by periodical population reductions and environmental fluctuations; at equilibrium competition appears and diversity would be reduced by the exclusion of species. The annual status of early successional species (1–15 years) confirms the first part of this hypothesis. The decrease in diversity after 15 years, when perennial species overrule the annual ones, and after 60 years at the forest stage, seems to confirm the second part of Huston's hypothesis. These decreasing diversity stages are linked with an increasing growth rate of one species: first *Brachypodium phoenicoides* and then *Quercus pubescens*. Their dominance is secured by an appropriate growth form, strategy of colonization and persistence in one community, i.e. perennial growth form, clump distribution, anemochory and barochory.

Moreover, some characteristics of vegetation dynamics in the study area may account for our pattern on diversity trends: 1. the absence of a spontaneous shrub stage – this stage exists but is produced by man-caused disturbances (Debussche et al. unpublished) – . 2. the absence of a *Pinus* stage with rapid colonization which would allow to pass gradually to the deciduous forest, and 3. the absence of numerous tree species which would prevent the exclusive dominance of *Quercus pubescens*.

We may remark that woody species belonging to more mature stages of succession colonize very early and contribute to the diversity, in the olive groves for example, and accelerate the succession. Therefore, late-successional species are able to settle without any preparation of the site by earlier ones (Drury & Nisbet 1973). Their establishment and growth to maturity do not require conditions produced by earlier species as demonstrated by Reiners et al. (1971).

The decrease in species turn-over rate during the succession (up to 20 year) (Fig. 3) is explained by the individual species characteristic in the various stages. In the beginning, most of the species are annuals which develop during spring and summer. The phytocoenosis thus regenerates annually and every species is subject to the risks of annual germination. When perennials settle, the area occupied by annuals decreases. When the canopy is closing the phytocoenosis is dominated by phanerophytes with a long life span, and species turn-over rate is low.

In the forest stages of the succession studied, *Quercus pubescens* is dominant in the canopy. Below 2 m the foliage distribution of the most abundant species is located in the same layers. Above the 2 m level, in La Rouvière-1, the foliage distribution of *Cornus sanguinea* along the vertical axis coincides with that of *Crataegus monogyna*, between 2 and 3 m. The joint foliage distribution of different species in these layers may be the result of the low growth-rate of species which established in the community at the same time. It is also possible that growth conditions within these layers are favourable, or that growth is limited to these layers because of the dominance of *Quercus pubescens* in the canopy.

In the understory of the two forest communities the horizontal pattern of species appears to result from heterogeneity in canopy density and hence in light conditions. Many authors demonstrated that forest trees impose a pattern of microsites on the forest floor (Moir 1966, Anderson et al. 1969, Bratton 1976 a, b, Hicks 1978). Struik & Curtis (1962), Smith & Cottam (1967), emphasized the importance of small-scale variability in forest soil properties by influencing species distribution on micro-topographic gradients.

Struik & Curtis (1969) and Horn (1971) identified the vegetative reproduction as the only cause of clumped distribution of many forest species. Mosaics of *Cornus sanguinea* in La Rouvière-1 and of *Festuca spadicea* and *Brachypodium pinnatum* in La Rouvière-2 may also result from vegetative reproduction. the asynchrony in reproduction between woody species, as noticed in La Rouvière-1, and between herbaceous species, in La Rouvière-2, was also observed by Mellinger & Mc Naughton (1975), Smith (1975), Newell et al. (1978). Recent studies stressed the competition between plants for pollinators (Levin & Anderson 1970, Mosquin 1971, Waser 1978) and the animal vectors of the seeds (Snow 1966). This competition seems to enhance phenological spreading (Mosquin 1971). All woody species in La Rouvière-1 are dispersed by zoochory. This may induce competition phenomenons between species for the dispersal agents, and the asynchronism in fruit dispersion may represent a way to limit competition, as noticed by Smith (1975). If we refer to the concept of guild (Root 1967), defined as coexisting populations utilizing the same class of resource in a similar way, we may distinguish several functional guilds, related to a resource utilization (such as light) or based on seasonal or stratal co-occurrence. In La Rouvière-1 for example, three groups of species coexist in a guild based on stratal co-occurrence: 1. *Cornus sanguinea*, *Prunus spinosa*, *Crataegus monogyna* and *Phillyrea angustifolia* between 20 and 50 cm. 2. *Prunus spinosa*, *Cornus sanguinea* and *Phillyrea angustifolia* between 1.40 and 1.50 m and 3. *Cornus sanguinea* and *Crataegus monogyna* between 2 and 3 m.

*Prunus spinosa*, *Cornus sanguinea* and *Crataegus monogyna* are separated during their flowering and fruit dispersal periods; *Cornus sanguinea* and *Phillyrea angustifolia* have a different response to light. (*Phillyrea angustifolia* was not considered phenologically, because there were not enough data available).

In La Rouvière-2, the two dominant species in the understory, *Brachypodium pinnatum* and *Festuca spadicea*, form two different guilds based on seasonal occurrence and difference in light resource utilization.

Coexistence within-functional guilds may partly depend on differences in the ecology of early life-history stages in fruit and seed morphology (Hicks, unpublished) but also in seasonally co-occurrence and in resource utilization (Bratton 1976a). This gives the possibility of a great species richness in communities, as we may conceive numerous combinations of the occurrence of two or more species in different functional guilds.

In the sites which we have studied, a small number of species combinations is to be found in functional guilds as a result of a reduced diversity, due to *Quercus pubescens* dominance in the canopy. However, diversity is maintained as species seem to be adapted to different parts of the vertical and horizontal gradient of light and of the annual climatic cycle. We may explain their coexistence in terms of spatial heterogeneity of the environment and of temporal separation of the functions as expressed by Grime (1978). This should represent a means to minimize the competition between species inside the community.

## Summary

Plant communities in the Mediterranean region have been analyzed with the emphasis on community processes regulating diversity trends. We want to know how several mechanisms that contribute to the maintenance or to the modification of species diversity, act: (i) in different communities, during an ecological succession (from old fields to forests), and (ii) in forest communities throughout the year. Some of these mechanisms were investigated through difference in stratification, life-form, phenology, dispersal agent, and microenvironmental gradient.

The paper consists of two parts. The first one deals with the relationship between structural and biological diversity in conjunction with succession stages leading to a *Quercus pubescens* forest. It is concluded that diversity, however it is measured, fluctuates during the succession: it increases initially (15 years after abandonment) and also in later stages when trees appear in the fields; it decreases after 20 years when *Brachypodium phoenicoïdes* dominates and in the latest stages of the succession studied. These trends may be interpreted in relation to the evolution of life form and dispersal agents spectra during the succession.

The second part concerns community-structure and biological diversity in relation to the organization of two *Quercus pubescens* forests. It appears that the species richness in the understory of the forests, may be explained by: (i) the tendency of the modal height of species foliage distribution to occupy different positions, (ii) the quasi-complete asynchronism in the major phenophases of the dominant species, (iii) different horizontal patterns of species distribution along a light gradient.

These schemes are discussed in relation to ecological hypotheses on diversity in relation to succession, and to community organization.

## References

Aber, J.D. 1979. A method for estimating foliage-height profiles in broadleaved forests. J. Ecol. 67: 35–40.

Anderson, R.C., O.L. Loucks & A.M. Swain. 1969. Herbaceous response to canopy cover, light intensity and throughfall precipitation in coniferous forests. Ecology 50: 255–263.

Auclair, A.M., & F.G. Goff. 1971. Diversity relations of upland forests in the Western Great Lakes area. Amer. Nat. 105: 499–528.

Bachacou, J. 1973. L'effet Guttman dans l'analyse de données phytosociologiques. I.N.R.A./C.N.R.F. Département de biométrie et de calcul automatique, Station de biométrie, Nancy, Document 73/5, 30 pp.

Bazzaz, F.A. 1968. Succession on abandoned fields in the Shawnee Southern Illinois. Ecology 49: 924–936.

Bazzaz, F.A. 1975. Plant species diversity in old field successional ecosystems in Southern Illinois. Ecology 56: 485–488.

Benzecri, J.P. et al. 1973. L'analyse des données 2. L'analyse des correspondances. Dunod, Paris, 619 pp.

Blondel, R. 1941. La végétation forestière de la région de St. Paul, près de Montpellier. Comm. S.I.G.M.A. no. 79.

Blondel, J. & R. Cuvillier 1977. Une méthode simple et rapide pour décrire les habitats d'oiseaux: le stratiscope. Oikos 29: 326–331.

Bratton, S.P. 1976a. Resource division in an understory herb community: responses to temporal and microtopographic gradients. Amer. Nat. 110: 679–693.

Bratton, S.P. 1976b. The response of understory herbs to soil depth gradients in high and low diversity communities. Bull. Torrey Bot. Club. 103: 165–172.

Braun-Blanquet, J. 1936. La chênaie d'Yeuse méditerranéenne. Comm. S.I.G.M.A. no. 40, Mémoires de la Société d'études des sciences naturelles de Nîmes, 14 pp.

Braun-Blanquet, J., J.M. Roussine, R. Nègre & L. Emberger. 1952. Les groupements végétaux de la France Méditerranéenne. C.N.R.S. et Direction de la Carte des Groupements Végétaux de l'Afrique du Nord. 297 pp.

Buell, M.F. & J.E. Cantlon. 1950. A study of two communities of the New Jersey Pine Barrens and a comparison of methods. Ecology 31: 567–586.

Canfield, R.H. 1941. Application of the line intercept method in sampling range vegetation. J. For. 39: 388–394.

Colinvaux, P.A. 1973. Introduction to ecology. Wiley, New York. 621 pp.

Connell, J.H. & R.O. Slatyer. 1977. Mechanisms of succession in natural communities and their role in community stability and organization. Amer. Nat. 111: 1119–1144.

Debussche, M., J. Escarré & J. Lepart. 1980. Changes in shrubs communities-examples of two mediterranean Papilionaceae. Vegetation (in press).

Dogniaux, R. & M. Lemoine. 1976. Programme de calcul des éclairements solaires énergétiques et lumineux de surfaces orientées et inclinées, ciel serein et ciel couvert. Inst. Roy. Météo. de Belgique. Miscellanea, Série C. no. 14, 16 pp.

Drury, W.H. & I.C.T. Nisbet. 1973. Succession. J. Arnold Arbor. 54: 331–368.

Ducrey, M. 1975. Utilisation des photographies hémisphériques pour le calcul des perméabilités des couverts forestiers au rayonnement solaire. I. Analyse théorique de l'interception. Ann. Sci. Forest. 32: 73–92.

Escarré Blanch, J. 1979. Etude de successions post-culturales dans les hautes garrigues du Montpelliérais. Thèse Montpellier, 134 pp. + ann.

Fernandez, R. 1978. Les peuplements de Chêne pubescent des Hautes Garrigues du Montpelliérais, étude dendrométrique et écologique. Mémoire Ecole Nationale des Ingénieurs des Travaux des Eaux et des Forêts. 41 pp.

Grime, J.P. 1977. Evidence for three primary strategies in plants and its relevance to ecological and evolutionary theory. Amer. Nat. 111: 1169–1194.

Grime, J.P. 1978. Interpretation of small scale patterns in the distribution of plant species in space and time. In: A.H.J. Freysen & J.W. Woldendorp (eds) Structure and functioning of plant populations, p. 101–124. North Holland Publ. Co, Amsterdam.

Grubb, P.J. 1977. The maintenance of species richness in plant communities: the importance of the regeneration niche. Biol. Rev. 52: 107–145.

Harcombe, P.A. & P.L. Marks. 1977. Understory structure of a mesic forest in Southeast Texas. Ecology 58: 1144–1151.

Hicks, D.J. 1978. A niche analysis of the herbaceous stratum of Cove forest. Bull. Ecol. Soc. Amer. 59: 78 (summary).

Hill, M.O. 1973a. Diversity and eveness: a unifying notation and its consequences. Ecology 54: 427–432.

Hill, M.O. 1973b. Reciprocal averaging: an eigenvector method of ordination. J. Ecol. 61: 237–249.

Horn, H.S. 1971. The adaptative geometry of trees. Princeton University Press, Princeton. 144 pp.

Horn, H.S. 1974. The ecology of secondary succession. Ann. Rev. Ecol. Syst. 5: 25–37.

Houssard, Cl. 1979. Etudes de la structure de quelques taillis de Chêne pubescent (Quercus pubescens Willd.). Examples pris dans la région des Hautes Garrigues du Montpelliérais. Thèse Montpellier, 184 pp.

Huston, M. 1979. A general hypothesis of species diversity. Amer. Nat. 113: 81–101.

Janzen, D.H. 1970. Herbivores and the number of tree species in tropical forests. Amer. Nat. 14: 501–528.

Krochmann, J. 1973. Quantities of illuminating engineering for day light. UNESCO Congres 'Le soleil au service de l'homme'. Paris 1973.

Levin, D.A. & W.W. Anderson. 1970. Competition for pollinators between simultaneously flowering species. Amer. Nat. 104: 455–467.

Lewontin, R.C. 1965. Selection for colonizing ability. In: H.G. Baker & G.L. Stebbins (eds.) The genetic of colonizing species. p. 79–94. Academic Press, New York.

Mac Arthur, R.H. 1969. Species packing and what interspecies competition minimizes. Proc. Nat. Acad. Sci. (U.S.) 64: 1369–71.

Mac Arthur, R.H. & H.S. Horn. 1969. Foliage profile by vertical measurements. Ecology 50: 802–804.

Mac Arthur, R.H. & E.O. Wilson. 1967. The theory of island biogeography. Princeton Univ. Press. Princeton. 203 pp.

Margalef, R. 1957. La teoria de la información en ecología. Mem. Real Acad. Cienc. Art. Barcelona 32: 373–449.

Margalef, R. 1968. Perspectives in ecological theory. Univ. Chicago Press. Chicago. 111 pp.

May, R.M. 1973. Stability and Complexity in Model Ecosystems. Princeton University Press, Princeton, 235 pp.

May, R.M. 1975. Pattern of Species Abundance and Diversity. In: M.L. Cody & J.M. Diamond (eds.) Ecology and Evolution of Communities, p. 81–120 Belknap, Harvard.

May, R.M. & R.H. Mac Arthur. 1972. Niche overlap as a function of environmental variability. Proc. Nat. Acad. Sci. (U.S.) 69: 1109–1113.

Mellinger, M.V. & S.J. Mc Naughton. 1975. Structure and function of successional vascular plant communities in Central New York. Ecol. Monog. 45: 161–182.

Moir, W.H. 1966. Influence of Ponderosa Pine on herbaceous vegetation. Ecology 47: 1045–1048.

Molinier, R. & P. Müller. 1938. La dissémination des espèces végétales. Rev. Gén. Bot. 50: 53–71 a.f.

Mosquin, T. 1971. Competition for pollinators as a stimulus for the evolution of flowering time. Oikos 22: 398–402.

Motomura, I. 1932. A statistical treatment of associations. Jap. Zool. 44: 379–383.

Murphy, G.I. 1968. Pattern in the life history and the environment Amer. Nat. 102: 390–404.

Newell, S.J. & E.J. Tramer. 1978. Reproductive strategies in herbaceous plant communities during succession. Ecology 59: 228–234.

Nicholson, S. & C.G. Monk. 1974. Plant species diversity in old field succession in the Georgia Piedmont. Ecology 55: 1075–1085.

Odum, E.P. 1969. The strategy of ecosystem development. Science 164: 262–270.

Pielou, E.C. 1969. An introduction to mathematical ecology. Wiley Interscience, New York 286 pp.

Pijl. L. van der. 1972. Principles of dispersal in higher plants (2nd ed.). Springer-Verlag, Berlin, Heidelberg, New York. 161 pp.

Reiners, W.A., I.A. Worley & D.B. Lawrence. 1971. Plant diversity in a chronosequence at Glacier Bay, Alaska. Ecology 52: 55–69.

Root, R.B. 1967. The niche exploitation pattern of the blue-gray gnat-catcher. Ecol. Monog. 37: 317–350.

Salisbury, E.J. 1942. The reproductive capacity of plants. Bell, London 244 pp.

Sharp, W.M. & V.G. Sprague. 1967. Flowering and fruiting in the White Oaks. Pistillate flowering, acorn development weather and yield. Ecology 48: 243–251.

Shugart, H.H. & J.M. Hett. 1973. Succession: similarities of species turnover rates. Science 180: 1379–1381.

Smith, A.J. 1975. Invasion and ecesis of bird-disseminated woody plants in a temperate forest sere. Ecology 56: 19–34.

Smith, B.E., & G. Cottam. 1967. Spatial relationships of mesic forest herbs in Southern Wisconsin. Ecology 48: 546–558.

Snow, D.W. 1966. A possible selective factor in the evolution of fruiting seasons in tropical forest. Oikos 15: 274–281.

Sokal, R.R. & F.A. Rohlf. 1969. Biometry. W.H. Freeman and Co. San Franscisco. 776 p.

Struik, G.J. & J.T. Curtis. 1962. Herb distribution in an Acer saccharum forest. Amer. Midl. Natur. 68: 285–296.

Vidal, J.P. 1979. Eléments historiques et bases phyto-écologiques pour une étude de la croissance du Chêne pubescent au bois de la Rouvière (Hérault) rapport de D.E.A., U.S.T.L. Montpellier 38 pp.

Waser, N.M. 1978. Competition for hummingbird pollination and sequential flowering in two Colorado Wildflowers. Ecology 59: 934–944.

Whittaker, R.H. 1965. Dominance and diversity in land plant communities. Science 147: 250–260.

Whittaker, R.H. 1969. Evolution of diversity in plant communities. Brookhaven Symposia in Biology 22: 178–196.

Whittaker, R.H. 1972. Evolution and measurement of species diversity. Taxon 21: 213–251.

Whittaker, R.H. 1975. Communities and ecosystems. (2nd. ed.). Mc Millan, New York. 385 pp.

Whittaker, R.H. 1977. Evolution of species diversity in land communities, In: W.C. Steere & B. Wallace (eds.) Evolutionary biology, vol. 10. p. 1–67 Plenum publishing Co, New York.

Whittaker, R.H. & S.A. Levin. 1977. The role mosaic phenomena in natural communities. Theor. Popul. Biol. 12: 117–139.

Yarranton, G.A. & R.G. Morrisson. 1974. Spatial dynamics of a primary succession: Nucleation. J. Ecol. 62: 417–428.

Accepted 15 December 1979

**Appendix**

Elucidation of the successional position of the research plots, being an elaboration of answer to questions on the paper.

In a 'synchronic' approach, some hypotheses have actually to be proved. The most important ones are: (i) the environmental homogeneity (climate, substratum, ...) of the studied fields, (ii) the same intensity and the same character of the disturbance preceding the abandonment (after a fire, one succession may differ from another one which has been brought about by different cultures) and (iii) the same development from the moment of abandonment to the date of the survey.

As indicated in the text the old fields are located in two different places (Le Causse de la Selle and La Boissière). In each of both places a few hundred meters apart, the climate and the substratum appeared to be homogenous enough to be able to confirm the hypotheses. In the case of the second hypothesis one has to consider that vine stocks sprout again a very long time after abandonment (from 20 to 25 years). The origin of the succession could be verified at least for the first stages. Moreover we must point out the possibility to verify the passing of a fire since abandonment, by reading the age-rings of some species such as *Juniperus oxycedrus*, *Genista scorpius*. As a matter of fact, some individuals of these species are not killed when a fire occurs, but they show deep marks. Thanks to this we could find 'lateral' successions, (Debussche et al. 1980 not described in this paper, with stages where *Genista scorpius* is dominant. Although this is not very clear, the reciprocal averaging method (Hill, 1973b) on a presence absence basis (Fig. 11), seems to show those 'lateral' successions. The phenomenons described in this article correspond to what we have thought to be 'main' succession, i.e. without any fire-caused disturbance.

As for the third hypothesis, the development of the fields from the moment of abandonment to the day of the observations is the same in our case. It is very difficult, however to verify the intensity of sheep grazing up to ten years before the observation. time. However, we may consider that all old fields have been grazed. The results (Fig. 11) of the reciprocal averaging method also appear to confirm our facing a single succession, without any sudden discontinuity in the species composition of the relevés. The variations of the turn-over rate have been reported in detail in the text.

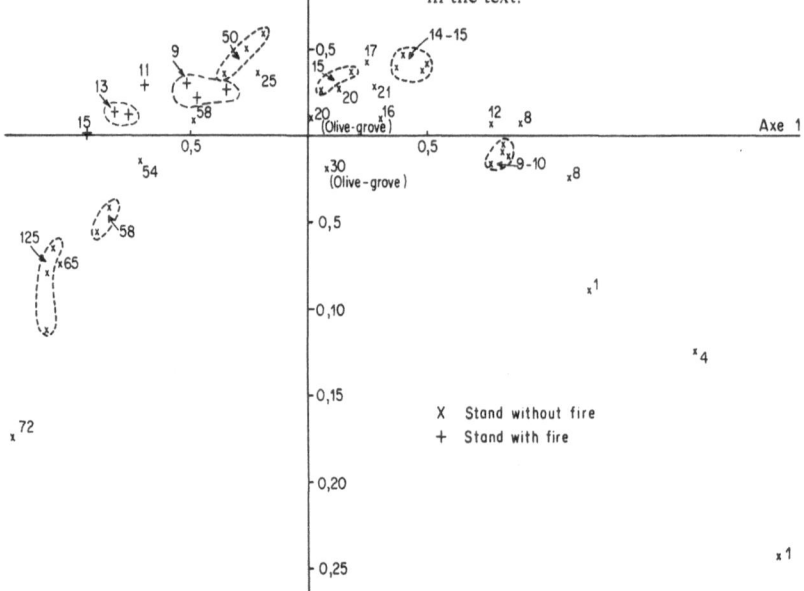

Fig. 11. 'Reciprocal averaging analysis' of stands, in the Montpellierais region (Hérault), belonging to a successional sequence, based on correlations between presences in each stand. Code numbers represent the ages of the relevés; dotted lines group relevés of the same age.

# CHANGES IN MEDITERRANEAN SHRUB COMMUNITIES WITH CYTISUS PURGANS AND GENISTA SCORPIUS*

M. DEBUSSCHE[1], J. ESCARRÉ[1], & J. LEPART[2]**

[1] Centre d'Etudes Phytosociologiques et Ecologiques Louis Emberger, C.N.R.S., B.P. 5051, 34033 Montpellier, Cedex France
[2] Ecothèque Méditerranéenne, C.N.R.S., B.P. 5051, 34033 Montpellier Cedex, France

**Keywords:**
Cytisus purgans, Dynamics, Genista scorpius, Languedoc, Mediterranean, Post-fire succession, Richness, Shrub community, Tree colonization

## Introduction

Shrub communities (rather than forests) are very important in the mediterranean landscape. The maintenance of these communities is primarily linked with human activities which have been carried out for centuries. Generally, the dynamics of the shrub communities seem to be very slow; their colonization and their succession by trees may be observed in scattered places.

These facts are related to biological characteristics of the dominant species, as for example: evergreen and dense foliage of leaves or 'twigs', a thick litter layer decaying very slowly, vigorous sprouts growing rapidly after burning or cutting, creeping roots with shoots, seeds germinating in great numbers on eroded or burnt spots, even after staying dormant for many years. For the dominant species, the combination of these strategies controls the patterns of changes of their communities in structure, richness, types of life forms and ability of colonization by trees.

## The studied communities in their regional context

On the southern side of the French 'Massif Central', near Montpellier, we investigated parts of two regions: south-wards, 'Les Garrigues', a mainly calcareous region of hills, small 'causses' (i.e. calcareous table-land and basins, and northwards, 'Les Cévennes', a schistous and granitic region of steep slopes, the highest point of which is Mont Aigoual (1565 m), 70 km from the Mediterranean Sea.

The precipitation gradient varies from 700 mm/yr at sea-level to 2250 mm/yr on the top of 'Les Cevennes'; the precipitation decreases rapidly on the northern side of Mont Aigoual through a 'föhn' effect. However, moisture conditions are generally intermediate during summer.

In the past centuries these regions were rather densely populated. The inhabitants lived on corn, vineyards, chestnut and sheep management. They cut down nearly all the forests. Nowadays after the 'rural exodus', woody species recolonize the landscape.

The majority of the widespread shrubs of physiognomic importance in these two regions are evergreens producing a slowly decaying litter, sprouting more or less vigorously after a fire or cutting. Examples are: *Quercus coccifera, Spartium junceum, Genista scorpius, Buxus sempervirens, Erica scoparia, Erica arborea, Erica cinerea, Calluna vulgaris, Sarothamnus scoparius, Cytisus purgans.* Creeping roots are particularly developed with *Quercus coccifera; Cistus monspeliensis* and *Ericaceae* show an explosion of seedlings after a fire; a long dormancy of seeds ('hard seeds') is found in *Papilionaceae.*

There are only a few studies on shrub dynamics in our regions, e.g. on *Quercus coccifera* settlements, by Trabaud (1980), Godron et al. (in press), Poissonet et al. (1978), Trabaud & Lepart (1980). In northern Europe, changes in *Calluna vulgaris* communities are rather well

---

*Plant nomenclature follows Fournier (1961); for birds see Peterson et al. (1967).
**We are very much indebted to Ch. Rimbault for her contribution to the translation of this paper.

known: Watt (1947), Heinemann (1956–57), Gimingham (1972, 1978), Barclay-Estrup (1970, 1971), Barclay-Estrup & Gimingham (1969), Noirfalise & Vanesse (1976). We also mention the studies about post-fire successions in the Californian chaparrals, e.g. Horton & Kraebel (1955), Patric & Hanes (1964), Hanes & Jones (1967), Mooney & Parsons (1973). We notice that some authors such as Niering & Egler (1955), Niering & Goodwin (1974), Webb et al. (1972), Spatz (1980), emphasized the role of shrubs to prevent the colonization by trees.

In this contribution changes in communities with *Cytisus purgans* and *Genista scorpius* (*Papilionaceae*) are analysed. *Cytisus purgans* grows on the misused ranges of 'Les Cévennes' region, usually between 600 and 1400 m elevation; it is known from the Centre and South of France, Spain, Portugal, Algeria and Morocco. *Genista scorpius* occurs especially on the old fields of 'Les Garrigues'; it occurs throughout the South of France, Spain and Portugal.

## Methods

Data were selected from two studies: (1) post-fire successions in *Cytisus purgans* communities of the northern side of Mont Aigoual (Debussche 1978) (2) post-cultural and post-fire successions, where *Genista scorpius* plays a rather prominent part, on a little causse in 'Les Garrigues' (Escarré 1979). For both studies, the approach of dynamics is the indirect one (Pavillard 1935).

Fifty stands were selected through stratified sampling among the *Cytisus purgans* communities. Each relevé includes:
– the states of 42 variables concerning localisation, topography, lithology, soil, and vegetation structure;
– the number of species and their cover values which were estimated by their presence on 128 points and segments along two 64 m parallel lines (each segment is 1 m long; the distance between the two lines is 4 m);
– vegetation structure, which was analyzed by counting the number of contacts of each species (and for shrubs each organ) with 10 cm intervals along a vertical pin; (64 samples per 20 cm along a line).
– the above-ground phytomass and litter, for each species (and for shrubs each organ) which was determined on a 4 m × 1 m rectangular plot; one of the long sides is next to the structure line;
– the age of the aerial parts of shrubs, (relative to the last fire), by counting the annual rings;

– the number, and the age (if possible), of the seedlings and young trees which grow on the 4 m × 64 m plot between the two lines.

The *Genista scorpius* community stands were selected according the age of the aerial parts growing after the fire (once again by counting the annual rings). The stands were clustered in a small area where ecological conditions are homogeneous. The data given here were collected in the following manner:
– richness was measured by counting the herbaceous species in 20 one $m^2$ plots and the woody species in a 200 $m^2$ plot;
– the structure was analysed by counting the presence per 10 cm stratum (in the present study called 'contact') of each species (and for shrubs each organ) along a vertical pin; 100 samples were taken each 20 cm along a line.

## Origin and spreading of the communities

To describe the stages of overgrowing by *Cytisus purgans* and *Genista scorpius*, we propose two schemes (Fig. 1, 2) which summarize the observations collected in situ and for which we also used the studies of Crocker & Barton (1957), Ellison (1960), King (1966), Côme (1970), Ahlgren (1974), Harper (1977), concerning equivalent processes or closely related species. However, some points should be examined by experiments, particularly to distinguish the precise role of fire and each factor breaking the dormancy.

Probably certain human activities increased some processes which had been present long before man had influenced the landscape: controlled and uncontrolled burnings taking the place of lightning fires, and sheep taking the place of wild herbivores. Prior to human activities both shrubs must have been confined to naturally burnt places and rocky outcrops.

Rangeland invasion by *Cytisus purgans* follows a period of overgrazing, as during the 18th and 19th century, (Dugrand & Werey 1971). Erosion together with overgrazing favour the seed dispersal and their burying. Sheep may carry the pods and disperse them. The seed coats are damaged by the (generally sandy and stony) moving top soil, caused by sheep trampling and restarting erosion. The seed coats are also probably softened by biochemical changes during ageing in the ground. *Cytisus purgans* which is very seldom grazed (only when the grass is wet), eliminates many of the weak remnant herbaceous species, and in the end the rangeland is abandoned by sheep. Fires do not kill the shrubs; a few days after the fire they are

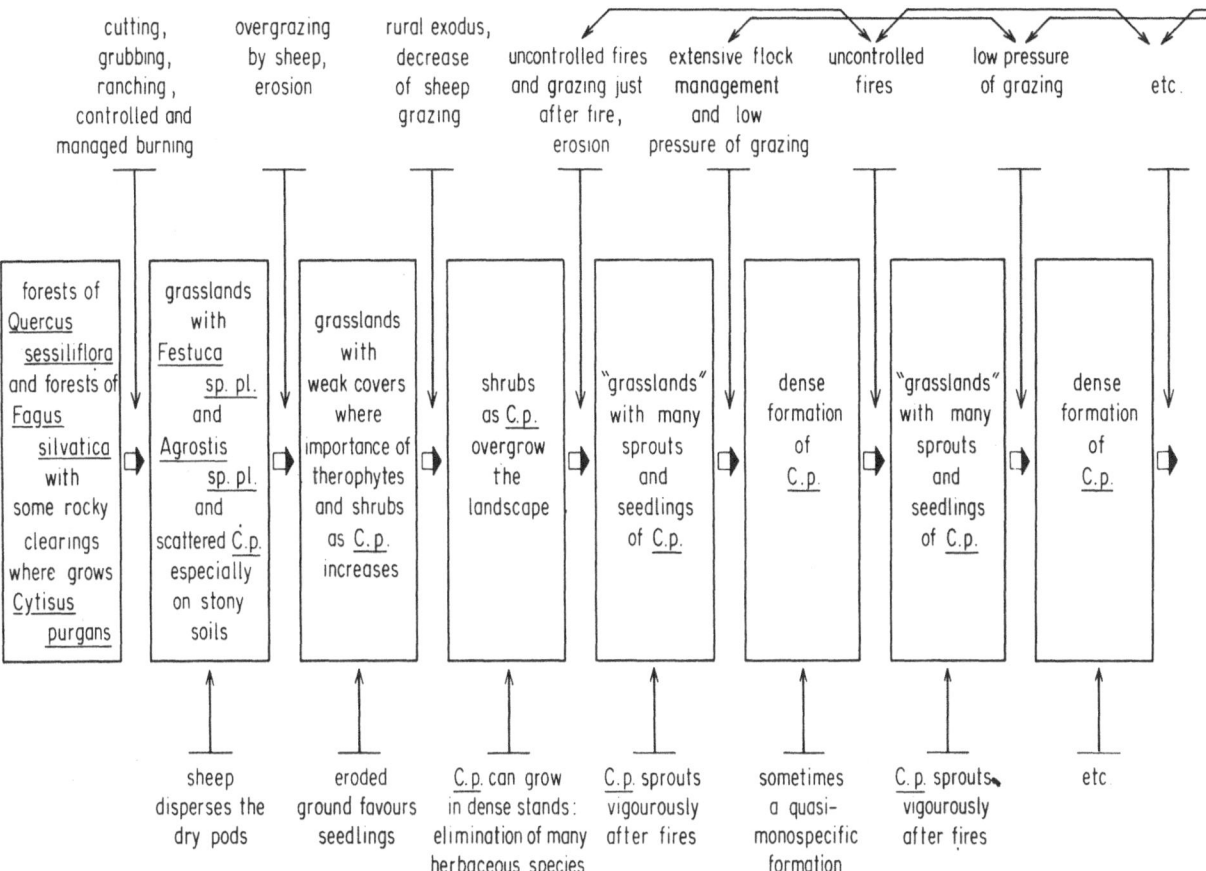

Top labels (inputs, above boxes):

cutting, grubbing, ranching, controlled and managed burning | overgrazing by sheep, erosion | rural exodus, decrease of sheep grazing | uncontrolled fires and grazing just after fire, erosion | extensive flock management and low pressure of grazing | uncontrolled fires | low pressure of grazing | etc.

Boxes (left to right):

forests of Quercus sessiliflora and forests of Fagus silvatica with some rocky clearings where grows Cytisus purgans | grasslands with Festuca sp. pl. and Agrostis sp. pl. and scattered C.p. especially on stony soils | grasslands with weak covers where importance of therophytes and shrubs as C.p. increases | shrubs as C.p. overgrow the landscape | "grasslands" with many sprouts and seedlings of C.p. | dense formation of C.p. | "grasslands" with many sprouts and seedlings of C.p. | dense formation of C.p.

Bottom labels (below boxes):

sheep disperses the dry pods | eroded ground favours seedlings | C.p. can grow in dense stands: elimination of many herbaceous species | C.p. sprouts vigorously after fires | sometimes a quasi-monospecific formation | C.p. sprouts vigorously after fires | etc.

Fig. 1. Origin and spreading of *Cytisus purgans* communities.

sprouting; moreover many seedlings appear after the passing of sheep. About 8 to 10 years after a fire, a compact canopy, more than one m high, has developed.

The invasion strategies by *Genista scorpius* are rather the same as those mentioned above; however, in this case fires seem to have a greater importance to initiate the growth of seedlings; the dispersal by sheep is, perhaps because of the thorny pod, more frequent. After a fire the sprouts are clearly less vigorous and the vegetation is generally less closed than in the case of *Cytisus purgans*.

**Changes in structure**

Both shrubs produce very small leaves which fall within a few weeks; so, the photosynthetic organs are 'green twigs', covered with scarce fuzz (*Cytisus purgans*) and hard thorns (*Genista scorpius*). These 'green twigs' are functional during two or three years and then dry out while remaining attached to the stems; they can stay like 'above-ground litter' for several years; mostly microclimatic conditions (wind and snow particularly) effect their fall. The light interception is achieved, not only by a photosynthetic stratum, but also by a sub-layer of dead matter; Baudière (1970) measured that in a dense stand of *Cytisus purgans*, only 0.7 % of the incidental light reached the ground, and Demaison (1976) found even smaller values.

Four examples of stratification at four different ages have been selected from all *Cytisus purgans* stands (Fig. 3); they enable us to emphasize the following remarks:
– the type of distribution of the chlorophyll – containing shrub stratum varies according to the age of the community; the greatest number of contacts (which in the present case can be more or less equalized with a maximum leaf index area) is located under the modal class in the young stands and above the modal class in the oldest ones.

75

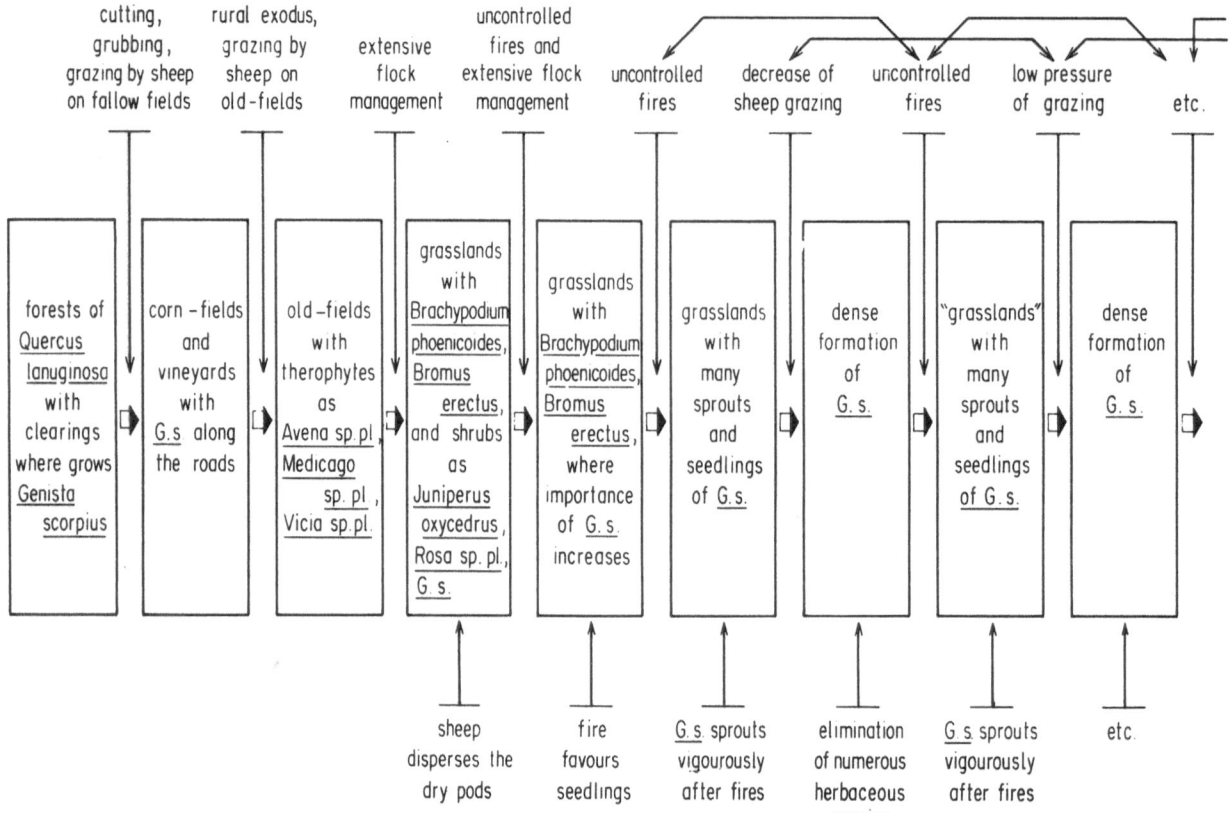

Fig. 2. Origin and spreading of *Genista scorpius* communities.

– the importance of herbaceous species decreases as the stand grows older;
– 'green twigs' and 'dry twigs' are found into two clearly distinct strata in the oldest stands.

The analysis and comparison of the 50 stands resulted in a theoretical graph, showing the development of the distribution of the chlorophyll-containing stratum in the shrub as well as a theoretical graph showing the development of the distribution of living and dead matter in the community (Fig. 4, 5); these graphs are related to communities with medium moisture conditions.

The graph of the 'green twigs' in the oldest stands is of the same type as the graph of the leaf index area for an old stand of *Quercus ilex* (Méthy 1974). Observations of some tree stands such as those of *Pinus pinea* confirm what was mentioned before: young trees have a relatively slender canopy whereas the old trees show needles in a flat dome. Debussche (1978) explains the shape of these distributions for the top of the graph from the decrease with age of the apical bud dominance and the shortening of the annual

growths, and for the lower part of the graph on the basis of the decrease of light throughout a rather homogeneous vegetal screen (Monsi & Saeki 1953, Chartier 1966, Horn 1971, Méthy 1974, Pianka 1978).

The study of the four examples taken from the *Genista scorpius* stands supports what was mentioned concerning *Cytisus purgans* (Fig. 6); however, the fourth diagram (18 yr) shows the situation during the senescent phase of the shrub:
– very few 'green twigs' remain but some 'green twigs' appear at the base of the stems (already visible on the third diagram and also noticed in some old *Cytisus purgans* stands.
– some sprouts dried out completely and some 'dry twigs' are to be seen above the 'green twigs';
– some sprouts collapsed (due to snow; this was also noticed for *Cytisus purgans*) while 'dry twigs' attached to the stems and 'dry twigs' making litter are continuously found;
– some species such as *Rubus ulmifolius* and *Prunus spinosa*

76

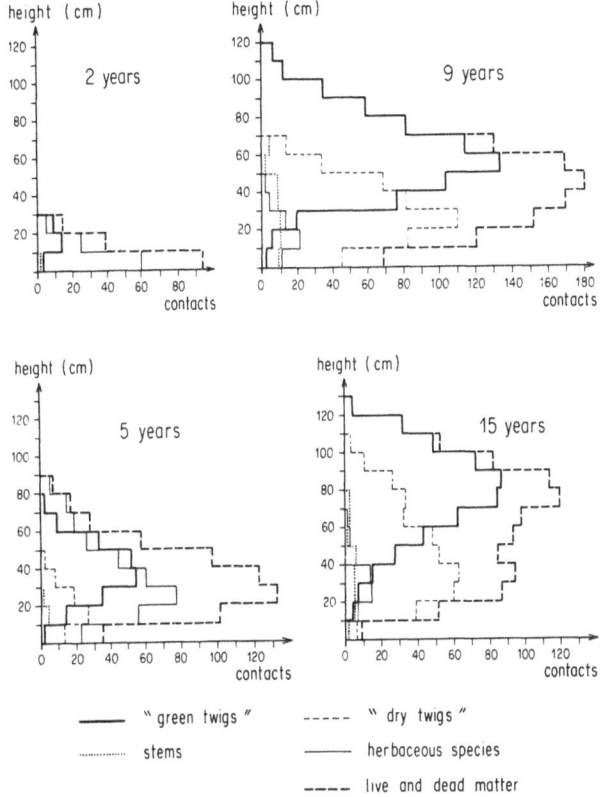

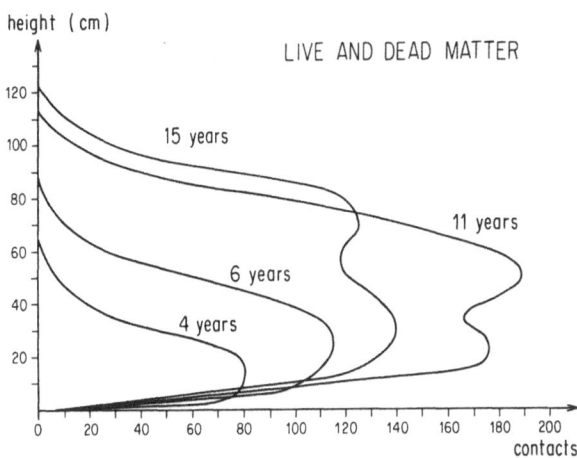

Fig. 5. Theoretical graph of development of *Cytisus purgans* community stratification (Debussche 1978).

Fig. 3. Four examples of stratification of *Cytisus purgans* communities (Debussche 1978).

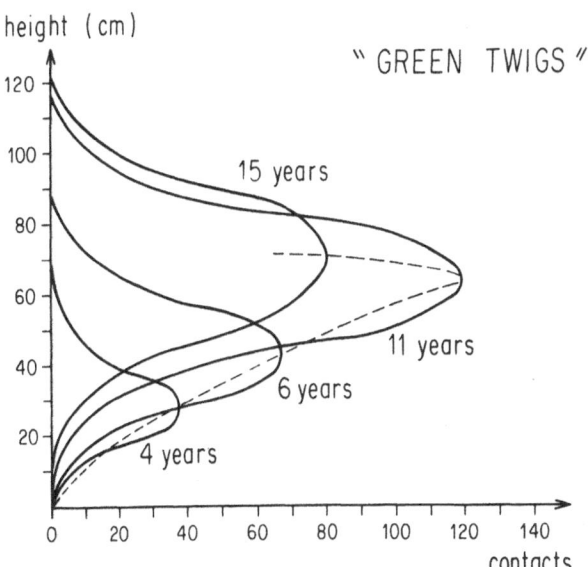

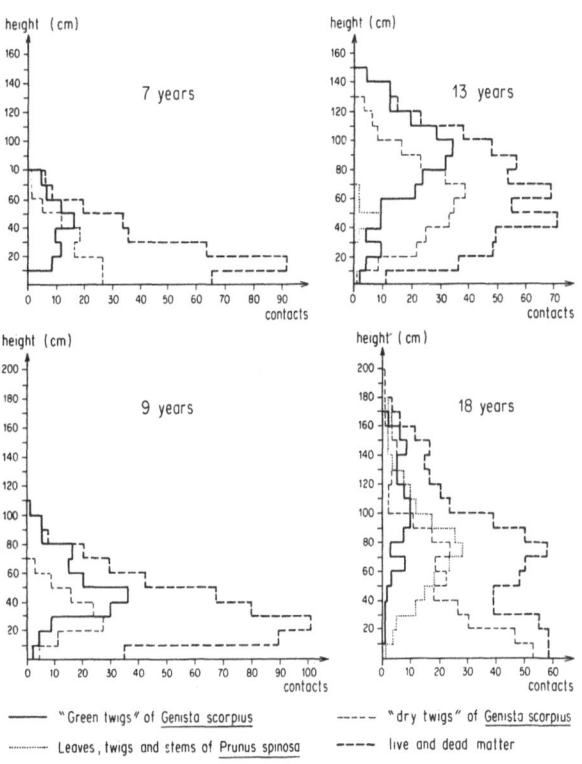

Fig. 4. Theoretical graph of development of *Cytisus purgans* 'green twigs' stratification (Debussche 1978).

Fig. 6. Four examples of stratification of *Genista scorpius* communities (Escarré 1979).

77

tend to gradually replace *Genista scorpius* (to simplify the diagram only *Prunus spinosa* is drawn here).

It is possible that some insects feeding on the shrubs and developing important populations for some years weaken them and promote the senescent phase (Demaison 1976), but this effect is not comparable to the attacks of *Locmaea suturalis* (L.) Thoms, and their effect on *Calluna vulgaris* stands (Heinemann 1956–57).

### Changes in richness and life forms

In our examples, changes in richness and life forms have been followed just after the fire for the *Cytisus purgans* stands and, from seven years after the fire, for the *Genista scorpius* stands; in the case of the *Cytisus purgans* stands, the shrub cover (live and dead matter) reaches 90 %– 100 %, here no gaps occur in the central part of the canopy; in the case of the *Genista scorpius* stands, the shrub might become senescent, but in that case the non-decayed litter still forms a thick cover.

There is a direct relation between changes in structure and changes in richness and life forms. Immediately after the fires, the perennial and fire resistant species which were present in the community before the fire appear again. Then they are accompanied by annuals and some bi-annuals and perennials, which survive as long as the shrubs do not overshadow them and the accumulation of litter does not suppress them (the maximum weight of litter measured is 9.0 kg of dry matter on 4 m² for a *Cytisus purgans* stand). When a gap occurs in the canopy a few species of *Rosaceae* such as *Prunus spinosa*, *Rosa sp. pl.*, *Rubus sp. pl.* develops in both shrub communities. When the gap in the canopy increases and the litter has been decomposed a number of new species appears.

Five successions, organized along a moisture gradient have been defined, throughout the 50 *Cytisus purgans* stands studied, on the basis of the mutual information technique combined with factor analysis (Debussche 1978). Only three of these five successions are considered here: the dryest one (succession A), the wettest one (succession C), and a medium one (succession B), which include one half of the stands (Fig. 7). A 'peak' of richness is evident in the case of succession A which occurs on steep, dry and rocky slopes; this maximum is generally reached during the second or third year after the fire; a rather similar graph is suggested by Shafi & Yarranton (1973) about burning of boreal forest stands and by Trabaud & Lepart (1980) in some woody mediterranean settlements. Al-

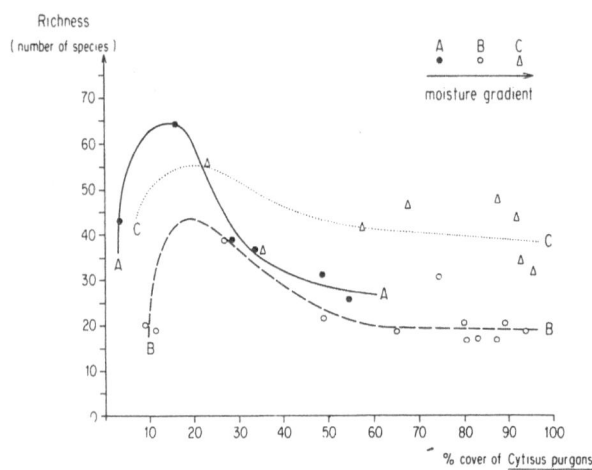

Fig. 7. Changes in richness along three post-fire successions in the *Cytisus purgans* communities.

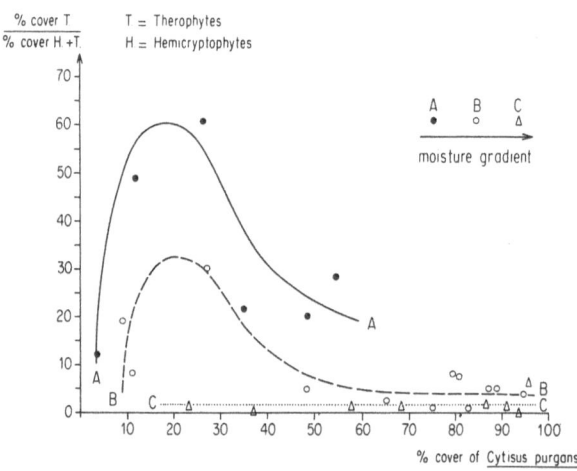

Fig. 8. Changes in the ratio between therophytes cover and hemicryptophytes cover along three post-fire successions in the *Cytisus purgans* communities.

though there is no relevé, available to show the low cover degree of *Cytisus purgans*, one can estimate that, in the case of succession C, the maximum is not pronounced and that there is broadly rather the same number of species whatever the cover is.

Changes in the ratio between therophyte and hemi-cryptophyte cover are interesting to follow (Fig. 8). In the case of succession A and succession B, the maximum of richness can be mainly ascribed to therophytes; the number of therophytes and their relatively important cover can be explained, on the one hand, by the dry environment (Cain 1950 Daget et al. 1978) and, on the

78

other hand, by the shelter role played by the rocks occurring in the stands. The role of the therophytes is unimportant in the case of succession C, which develops on wet soils where some perennial herbaceous species, both fire resistant and shade tolerant, re-colonize quickly the bare ground.

In the case of *Genista scorpius* communities, as in the cases of succession A and succession B, there is a very clear decrease of richness with age (Fig. 9); the therophytes still remain nine years after the fire and then all disappear (similarity with succession B), whereas the hemicryptophytes remain as relicts 18 yr after the fire.

## Colonization by trees

To achieve a colonization by trees, it is obvious that se have to reach the ground and germinate. The seedlings may develop into trees, which gradually replace the shrubs by overshadowing. The abundance of the seed rain on a stand depends on the distance between the trees and the stand, on the number of trees, on the types of dispersal used by the trees, on the quality of each vector and on the quality of each tree as a dispersal source (Harper 1977). When the seeds fall on the ground, many of them are eaten by rodents, birds, or insects, or they are destroyed by fungi and parasites; the rest, (depending on the species), can either lay dormant, or germinate if they are in favourable sites, or disappear if they are not; 'unfortunately', the seeds of trees have generally short lives in the soil (Harper 1977). Then the seedlings must escape the grazing sheep and benefit from favourable growth conditions to become a tree. Thus, the success is small in comparison with the quantity of seeds produced.

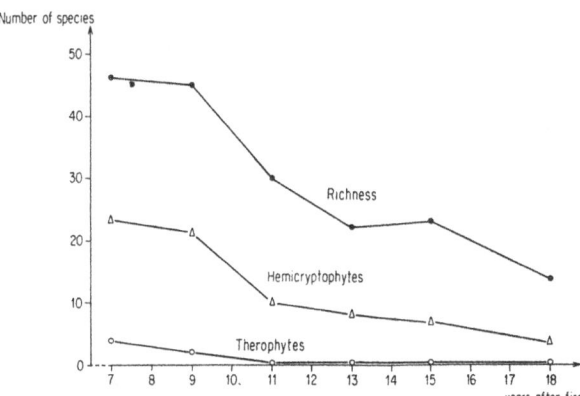

Fig. 9. Changes in richness, number of hemicryptophytes and therophytes along a post-fire succession in the *Genista scorpius* communities (Escarré 1979).

Therefore, the colonization by trees needs several conditions which are either dependent on the shrub community itself (internal conditions), or on the plant-matrix surrounding the shrub community patch and on animals (external conditions); man, too, plays an important role (by setting fires, by reafforesting, by sheep managing, etc.).

*Internal conditions*

The internal conditions which prevent the colonization by trees are in our cases:
– the density of the 'green twigs' and 'dry twigs' which strongly reduces the light intensity at the soil surface, prevents the colonization by trees with light dormancy-breaking seeds and etiolates the seedlings;
– a thick litter which buries the seeds and also prevents the colonization by trees with light dormancy-breaking seeds;
– the rapid alternations between moisture-dryness in the litter which destroy seeds and seedlings;
– the competition for water between seedlings, young trees and shrubs.
– perhaps toxic actions of the shrubs themselves and of the litter against seeds and seedlings.

Finally, we must point out the inflammability of these stands, each fire eliminating the pioneer trees, inasmuch as they are not able to sprout (either because of their species characteristics, or their youth, or the intensity of the fire).

*External conditions*

The trees which over a certain distance give the most important seed rain are trees with winged or plumed seeds This distance however, is rather short; a study on the seed rain of an american winged-seed *Pinus echinata* shows that 85 % of the seeds fall within 50 m of the forest edge (Yocom, 1968). In the study area of *Cytisus purgans* communities, several species of trees with winged seeds occur, but many of them are scattered individuals or small clumps, depending on a specific character of distribution (*Ulmus campestris* and *Acer campestre*), or on a biogeographical feature (*Betula verrucosa*), or on both of them (*Acer platanoides*). Introduced coniferous species, such as *Picea excelsa* and *Pinus laricio*, are found in localized reafforestations. Further, only *Fraxinus excelsior*, occurring along streams, and *Pinus sylvestris*, are found in rather important stands. In the study area of *Genista scorpius* communities, *Ulmus campestris*, *Fraxinus oxyphylla*, *Acer campestre*, and *Acer monspessulanum* are the only winged-seed species of trees, but they are all rare, the first two occurring generally on wetter spots.

Trees with fleshy fruits, especially *Sorbus aria* and *Sorbus aucuparia* grow isolated or in small groups on the northern side of Mont Aigoual; *Sorbus domestica* and *Sorbus torminalis* are rare on the small *causse* where the study was done. The seeds of these trees are mainly dispersed by birds in late summer, autumn and winter, according to the ripeness of the fruit; thus, both migratory and resident birds participate in the dispersal, especially *Turdinaea*: *Turdus merula* (nesting and migratory populations); *Turdus philomelos* and *Turdus viscivorus* (nesting populations in 'Les Cévennes' region and migratory populations); *Turdus iliacus* and *Turdus pilaris* (migratory populations); *Turdus torquatus* occurs some years with important migratory populations on Mont Aigoual where very few pairs nest (Blondel et al., 1979; Lhéritier, in press). The homogeneous cover of the shrubs does not attract birds, and very few species nest here; for example, only seven bird species, not especially frugivorous, nest in the dense mixed *Cytisus purgans-Calluna vulgaris* stands at the northern side of Mont Aigoual, while 29 species nest in a *Pinus sylvestris* and *Fagus sylvatica* forest of the same area (Blondel et al., 1979).

Acorns, beech-nuts and chestnuts are dispersed either by the effect of gravity alone, or by the hiding habits of some birds, such as *Garrulus glandarius* (Chettleburgh 1952, Turček 1961), and some rodents, or carried away by water and deposited along the banks. Small mammals are very few in the dense shrublands; they concentrate on boundaries and edges (Treussier 1975). *Fagus sylvatica*, in groups, *Quercus sessiliflora* and *Quercus lanuginosa* (with hybrids), and *Castanea sativa* (which is rare) occur at different altitudes, more or less next to the *Cytisus purgans* stands. *Quercus lanuginosa*, in groups and some individuals of *Quercus ilex* grow next to the *Genista scorpius* stands.

The pattern in faciès of vegetation (i.e. of units with an own structure and dominant species sense Godron & Lepart 1975) is very important for the dynamics of each patch (Forman, 1980); in our two study areas, this pattern is quite different. *Cytisus purgans* generally grows in big patches (up to 30 or 40 ha) on the slopes, the monotony of which is generally interrupted by groups of *Pteridium aquilinum* and *Sarothamnus scoparius*, or by a line of *Fraxinus excelsior* which escaped fire. The surrounding plant formations are forests as well as shrublands, and more rarely crops. *Genista scorpius*, generally covers not more than one ha; it overgrows old-fields which show, together with vineyards a mosaic of different successional

stages around a half-deserted hamlet. The dividing walls and some former abandoned the old-fields are occupied by *Quercus lanuginosa*.

*Field observations*

Among the 50 *Cytisus purgans* stands studied, 32 have tree species; in 3 relevés, we observed sprouting (*Fagus sylvatica* and *Fraxinus excelsior*), in 16 relevés, we saw young trees with rather the same age as the shrubs and in 22 relevés, we observed seedlings or young trees quite younger than the shrubs. *Fraxinus excelsior* and *Pinus sylvestris* are the most common tree species.

The individuals of *Fraxinus excelsior* are rarely as old as the shrubs (2 relevés) but usually they are quite younger (15 rel.). The seeds of *Fraxinus excelsior* germinate and give seedlings mostly during the senescent phase of the shrub.

The individuals of *Pinus sylvestris* are as old as the shrubs (10 relevés) or quite younger, in stands older than fourteen years (6 rel.) The seeds of *Pinus sylvestris* only germinate either just after the fire when the ground is bare, or on places where stems have collapsed and no undecayed litter occurs.

The colonization by *Quercus lanuginosa* in the study area of the *Genista scorpius* community is quite evident: except in the dense grasslands of *Brachypodium phoenicoides* and in some dense stands of *Genista scorpius*; the seedlings appear almost in every plant formation, even a cultivated vineyard. However, sheep grazing hinders the colonization by injuring the young trees.

Therefore, the internal and external conditions required for colonization selected the tree species to appear and the periods to succeed, and they also induce the evidence and speed of overgrowing.

**Discussion and conclusion**

Clear similarities could be indicated in the changes of the two shrub communities of *Cytisus purgans* and *Genista scorpius*. For the colonization by trees both structure of the shrub stand and faciès pattern in the vegetation are important.

The observations on the changes in shrub communities exposed here fit with the 'inhibition' model proposed by Connell & Slatyer (1977). However, some other observed successional stages fit their 'facilitation' and 'tolerance' models. Thus, below 1,000 m altitude, on the northern side of Mont Aigoual, *Fagus sylvatica* will usually not be able to settle directly within grassland or shrub stands, whatever their age might be; it will have to wait for such tree species such as *Pinus silvestris* or *Fraxinus excelsior* modifying microclimatic conditions. If a species development fits the 'inhibition' model, species richness (and other diversity indices) usually have a low value, which will increase after such a species disappears. Since this phenomenon can be found in both herbaceous species, shrubs, and trees, the succession as a whole shows fluctuating values of richness, as suggested by Whittaker (1975).

The study of plant succession must be undertaken at complementary levels: at the level of the replacement of an individual by another individual, at the level of the community, but also at the level of the inter-relations between the studied community and the surrounding communities. Furthermore it is necessary to include the influence of man and animals which often play a prominent role in plant successions, either by promoting the arrival and subsequent spreading of species, or by promoting the decrease and subsequent elimination of others.

## Summary

Shrub communities are important in the mediterranean landscape and linked very much with human activities. The permanency and the spreading of these communities are related to biological characteristics of the dominant species which control the patterns of changes in structure, richness, 'types of life forms, and ability of colonization by trees. In this paper, changes in communities are analysed and compared for *Cytisus purgans* and *Genista scorpius* (both *Papilionaceae*). For both communities the schemes of origin and spreading are rather the same: overgrazing and fire play a prominent role. Changes in structure show different stages emphasized by the types of distribution of the chlorophyll containing stratum. There is a direct relation between changes in structure and changes in species richness and life form spectrum: richness decreases generally from the second or third year after fire. According to the dryness of the stand therophytes may or may not be important in the youngest stands. The colonization of the shrub communities by trees is dependent on several conditions, either on the shrub community itself, or on the plant-matrix surrounding the shrub community patch and also on animals; man, too, plays an important role. In conclusion we think that the above-mentioned changes in shrub communities fit the 'inhibition' model proposed by Connell & Slatyer (1977);

but, more generally it seems that the plant successions observed in our regions follow both, this model and the 'facilitation' and 'tolerance' models of the same authors. Linked with this fact, it seems that the successions show generally fluctuating values of richness as is suggested by Whittaker (1975). We think that plant succession must be well analyzed at different levels of organization and include the study of the influences man and animals may have.

## References

Ahlgren, C.E. 1974. Effects of fires on temperate forests: North Central United States in: T.T. Kozlowsky & C.E. Ahlgren (eds.) Fire and ecosystems 6: 195–219 ed. Academic Press, New-York, San Francisco, London.

Barclay-Estrup, P. 1970. The description and interpretation of cyclical processes in a heath community, II, Changes in biomass and shoot production during the Calluna cycle. J. Ecol. 58: 243–249.

Barclay-Estrup, P. 1971. The description and interpretation of cyclical processes in a heath community, III, Microclimate in relation to the Calluna cycle. J. Ecol. 59: 153–166.

Barclay-Estrup, P., & C.H. Gimingham. 1969. The description and interpretation of cyclical processes in a heath community, I, Vegetational changes in relation to the Calluna cycle. J. Ecol. 57: 737–758.

Baudière, A. 1970. Recherches phytogéographiques sur la bordure méridionale du Massif Central français (les Monts de l'Espinouze). Thèse, Fac. Sci., Montpellier, 3 vol., 317 p. + 567 p. + annexes.

Blondel, J. et al. 1979. Structure et dynamique des peuplements d'Oiseaux dans un secteur de moyenne altitude du Parc National des Cévennes: le versant nord-ouest de l'Aigoual. Ann. Sci. Parc Nat. Cévennes 1: 59–83.

Cain, S. 1950. Life-forms and phytoclimate. Bot. Rev. 16: 1–32.

Chartier, P. 1966. Etude du microclimat lumineux dans la végétation. Ann. Agron. 17: 571–602.

Chettleburgh, M.R. 1952. Observations on the collection and burial of acorns by Jays in Hainault Forest. British Birds. 45: 359–364.

Côme, D. 1970. Les obstacles à la germination. Masson et Cie, Paris. 162 pp.

Connell, J.H. & R.O. Slatyer. 1977. Mechanisms of succession in natural communities and their role in community stability and organization. Amer. Natur. 111: 1119–1144.

Crocker, W. & L.V. Barton. 1957. Physiology of seeds. An introduction to the experimental study of seed and germination problems. Chronica Botanica Company, Waltham. 267 pp.

Daget, Ph., J., Poissonet. & P. Poissonet. 1978. Le statut thérophytique des pelouses méditerranéennes du Languedoc in: J.-M. Géhu (ed.) La végétation des pelouses sèches à Thérophytes p. 81–99. J. Cramer, Vaduz.

Debussche, M. 1978. Etude de la dynamique de la végétation sur le versant nord-ouest du Mont Aigoual. Thèse doct.-ing., Univ. Sci. Tech. Languedoc, Montpellier, 2 vol.: 74 p. + 67 pp. + cartes.

Demaison, A. 1976. Contribution à l'étude écologique du peuplement arthropodien frondicole des landes à Genêt purgatif (Cytisus purgans (L.) Benth.) dans les Cévennes. Thèse doct. spéc., Univ. Sci. Tech. Languedoc, Montpellier, 108 p.

Dugrand, R. & G. Werey. 1971. Agriculture in: R. Dugrand (ed.): Atlas Régional du Languedoc-Roussillon. Berger-Levrault, Paris.

Ellison, L. 1960. Influence of grazing on plant succession of rangelands. Botan. Rev. 26, 1: 1–78.

Escarré, J. 1979. Etude des successions végétales dans la séquence à Quercus lanuginosa Lamk des Garrigues du Montpelliérais. Thèse doct. spéc., Univ. Sci. Tech. Languedoc, Montpellier.

Forman, R.T.T. 1979. The Pine Barrens of New Jersey, an ecological mosaic in: R.T.T. Forman in: Pine Barrens. Ecosystem and Landscape (ed): Academic Press, New York.

Fournier, P. 1961. Les quatre flores de la France, Corse comprise (générale, alpine, méditerranéenne, littorale). 1105 pp. Lechevalier, Paris.

Gimingham, C.H. 1972. Ecology of heathlands. Chapman & Hall, London. 266 pp.

Gimingham, C.H. 1978. Calluna and its associated species: some aspects of co-existence in communities. Vegetatio 36: 179–186.

Godron, M. & J. Lepart. 1975. Sur la représentation de la dynamique de la végétation au moyen de matrices de succession in: W. Schmidt (ed.) Sukzessionsforschung: p. 269–287. Cramer, Vaduz.

Godron, M., et al. (in press). Dynamique de la végétation et aménagement in: F. di Castri (ed.) Maquis et Chaparrals, V.

Hanes, T.L. & H.W. Jones. 1967. Postfire chaparral succession in Southern California. Ecology 48: 259–264.

Harper, J.L. 1977. Population Biology of Plants. Academic Press, London, New-York, San Francisco. 892 pp.

Heinemann, P. 1956–57. Les Landes à Calluna du district Picardo-Brabançon de Belgique. Vegetatio 2: 99–147.

Horn, H.S. 1971. The adaptative geometry of trees. Princeton University Press, Princeton. 144 pp.

Horton, J.S. & Kraebel, C.J. 1955. Development of vegetation after fire in the chamise chaparral of Southern California. Ecology 36: 244–262.

King, L.J. 1966. Weeds of the world. Biology and control. Leonard Hill, London, New-York. 526 pp.

Lhéritier, J.N. (in press). L'avifaune des Garrigues. Naturalia Monspeliensia.

Méthy, M. 1974. Interception du rayonnement solaire par différents types de végétation dans la région méditerranéenne. Thèse doct. Univ. Sci. Nat., Univ. Sci. Tech. Languedoc, Montpellier, 51 pp. + annexes.

Monsi, M. & T. Saeki. 1953. Über den Lichtfaktor in den Pflanzengesellschaften und seine Bedeutung für die Stoffproduktion. Jap. J. Botany 14: 22–52.

Mooney, H.A. & D.J. Parson. 1973. Structures and function of the Californian chaparral. An example from San Dimas in: F. di Castri & H.A. Mooney (eds.) Ecological studies, 7,

Mediterranean type ecosystems, Origin and structure, III, 1: 83–112. Berlin, Heidelberg, New-York.

Niering, W.A. & F.E. Egler. 1955. A shrub community of Viburnum lentago, stable for twenty five years. Ecology 36: 356–360.

Niering, W.A. & B.H. Goodwin. 1974. Creation of relatively stable shrublands with herbicides: arresting 'succession' on rights-of-way and pastureland. Ecology 55: 784–795.

Noirfalise, A. & R. Vanesse. 1976. Les landes à bruyère de l'Europe occidentale. 54 pp. (Conseil de l'Europe édit.) Strasbourg.

Patric, J.H. & T.L. Hanes. 1964. Chaparral succession in a San Gabriel mountain area in California. Ecology 45: 353–360.

Pavillard, J. 1935. Eléments de sociologie végétale. Act. scient. industr. 251: 102 p. Hermann et Cie, Paris.

Peterson, R., G. Mountfort. & P.A.D. Hollom. 1967. Guide des Oiseaux d'Europe. 447 p. 4th ed. Delachaux et Niestlé, Neuchâtel.

Pianka, E.R. 1978. Evolutionary Ecology. 397 pp. 2nd ed. Harper & Row, New-York, Hagerstown, San Francisco, London.

Poissonet, P., F., Romane, M. Thiault & L. Trabaud. 1978. Evolution d'une garrigue de Quercus coccifera L. soumise à divers traitements: quelques résultats des cinq premières années. Vegetatio 38: 135–142.

Shafi, M.I. & G.A. Yarranton. 1973. Diversity, floristic richness and species evenness during a secondary (post-fire) succession. Ecology 54: 897–902.

Spatz, G. 1980. Succession patterns on mountain pastures. In E. van der Maarel (ed.) Advances in vegetation science, B. Succession. Vegetatio 43: 000–000.

Trabaud, L. 1970. Quelques valeurs et observations sur la phytodynamique des surfaces incendiées dans le Bas-Languedoc. Naturalia Monspeliensia, 21: 231–242.

Trabaud, L. & J. Lepart. 1980. Diversity and stability in Garrigue ecosystems after fire. In: E. van der Maarel (ed.) Advances in vegetation science, B. Succession. Vegetatio 43: 000–000.

Treussier, M. 1975. Contribution à l'étude du peuplement micro-mammalien de l'Aigoual et des Causses. Thèse doct. spéc., Univ. Sci. Tech. Languedoc, Montpellier, 174 pp. + annexes.

Turček, F.J. 1961. Ökologische Beziehungen der Vögel und Gehölze. Verl. Slow. Akad. Wiss., Bratislava.

Watt, A.S. 1947. Pattern and process in the plant community. J. Ecol. 35: 1–22.

Webb, L.J., J.G. Tracey & W.T. Williams. 1972. Regeneration and pattern in the subtropical rain forest. J. Ecol. 60: 675–695.

Whittaker, R.H. 1975. Functional aspects of succession in deciduous forests in: W. Schmidt (ed.): Sukzessionsforschung p. 377–405. Cramer, Vaduz.

Yocom, H.A. 1968. Short leaf pine seed dispersal. J. Forest. 66: 422.

Accepted 20 December 1979

# PHENOLOGICAL SPREAD IN PLANTS: A RESULT OF ADAPTATIONS TO ENVIRONMENTAL STOCHASTICITY?

Torbjörn FAGERSTRÖM & Göran I. ÅGREN*

Swedish Coniferous Forest Project, Swedish University of Agricultural Sciences, S-750 07 Uppsala, Sweden

**Keywords:**
Adaptation, Colonization, Environmental stochasticity, Phenology, Weather variation

## Introduction

Plant species co-inhabiting a given geographical region often have distinctly different times of flowering. Such phenological spread is known from taxa pollinated by insects (Mosquin 1971), birds (Stiles 1975), and wind (Wells 1972, Pemadasa & Lovell 1974). With regard to the selective forces that promote the evolution of separate flowering times, two mechanisms have been suggested: (i) competition for pollinators, i.e. one plant species reducing rates of visitation to others by being more attractive to pollinators (Mosquin 1971, Stiles 1975), and (ii) competition for pollination, i.e. fecundity costs arising from a loss of pollen or stigmatic surfaces through interspecific pollinations (Levin & Anderson 1970, Waser 1978).

In qualitative terms, these two hypotheses are based on the assumption that an increased number of ovules being fertilized leads to an increased number of new seeds being produced and dispersed, an assumption that is likely to hold generally. In quantitative terms, however, they also require the further assumption that the marginal increase in fitness gained from producing an extra seed is significant. That assumption is less likely to hold generally, because in many communities the plants are facing intense inter- and intraspecific competition at the stage of seedling establishment. Then, for an inferior competitor the best strategy may be one that relaxes this competi-

tion, rather than one that strengthens it, i.e. producing more seeds.

In addition to such cases where the significance of the two mechanisms above can be questioned on quantitative grounds, there are cases where they may even be questioned à priori. For example, in the Polish *Tilia-Carpinus* forest studied by Falinska (1972, cited in Harper 1977) many of the species have so specialized flowering biologies that the incidence of interspecific pollination should be a rare event and competition for pollinators seems unlikely. The only conceivable interaction seems to be at the seedling stage. Yet, there is distinct differentiation with respect to flowering times.

We advance here the hypothesis that phenological spread in flowering times is a result of competition at the stage of seedling establishment. It is not intuitively obvious that segregation of flowering times may lead to stable coexistence rather than merely slowing down the extinction of an inferior species in communities with intense competition at the stage of seedling establishment. A hypothetical but extreme example might help understanding these questions. Consider therefore two perennial species which produce seeds only every second year and which have the same (constant) seed production per individual. Let further one of the species be superior at the stage of seedling establishment which is assumed to be completed within one year. It is clear that if the two species produce seeds in the same years, the superior species will eventually outcompete the inferior one. On the other hand, if the two species produce seeds alternate years the question of superiority or inferiority looses its meaning and stable coexistence can prevail.

In the present contribution we will analyse how much two species can approach each other phenologically

* We thank L.-E. Liljelund, C. Solbreck and C. Wiklund for helpful comments. This work was carried out within the Swedish Coniferous Forest Project, supported by the Swedish Natural Science Research Council, the Swedish Environmental Protection Board, the Swedish Council of Forestry and Agricultural Research, and the Wallenberg foundation.

with maintained coexistence. Formal proof is provided that for two species with simultaneous germination of seeds, competition for colonization space is reduced if the flowering times are separated. The fundamental element of our hypothesis is that environmental temporal stochasticity generates random variations in seed production. Thus, while the previous hypotheses focus only on the average seed production, we also include the variations in seed production.

## Theory

Consider an environment consisting of a fixed number of sites, each site supporting only one adult plant. Let this environment be inhabited by two plant species that reproduce only by dispersing seeds once a year. Let furthermore species 1 be a superior competitor in the seedling stage in the sense that a seed of species 2 that falls at a given empty site will develop into an adult if and only if there are no seeds of species 1 present. Finally, let the two species be identical with respect to viability of seeds, time of seed germination etc., such that the only factor determining the probability that a given empty site will be colonized by species 2 in a given year is the probability of that site having received at least one seed of species 2 and no seed of species 1 in that year.

At the time of seed dispersal in a given year a certain fraction of the total number of sites is available for colonization, viz. $\sum_{i=0}^{2} \lambda_i n_i(t)$. where $n_i(t)(i = 0, 1, 2)$ is the fraction of sites colonized by species $i$ in the year $t$, and $\lambda_i$ is the (constant) death rate of the adults of species $i$, regarding empty sites as colonized by species $0$ with death rate unity. Then, the fraction of sites available for colonization created through the death of specimens of species $i$ is

$$\frac{\lambda_i n_i(t)}{\sum \lambda_j n_j(t)} \qquad i = 0, 1, 2$$

Let $P_i(t)$ denote the probability that a given empty site is colonized by species $i$. For a large system, the temporal average $\langle P_i(t) \rangle$ then equals the fraction of empty sites colonized by species $i$. As a criterium for infinite coexistence we require the existence of positive solutions to the following system of equations

$$\langle P_i(t) \rangle = \frac{\lambda_i n_i^*}{\sum \lambda_j n_j^*} \qquad i = 0, 1, 2 \qquad (1)$$

where the superscript '*' denotes the stationary value.

We have elsewhere (Fagerström & Ågren 1979) derived the solutions to Eqs (1) and analyzed their stability properties, for the case where (i) the number of viable seeds produced per individual of species $i$ a given year is a Gaussian stochastic variable with mathematical expectation $a_i$, variance $\sigma_i^2$, auto-correlation zero and cross-correlation $\rho$ ($i = 1, 2$), and (ii) the seeds are dispersed at random over the entire system. Here the interest is focused on the qualitative results obtained [which should be insensitive to the specific assumptions (i) and (ii)]; more specifically, does it exist values of the parameter $\rho$ permitting species 2 (the inferior competitor) to survive, given that it is identical with species 1 in all other respects (i.e. $a_1 = a_2 = a$; $\lambda_1 = \lambda_2 = \lambda$; $\sigma_1 = \sigma_2 = \sigma$)? The condition for survival of species 2 can then be shown to be (Fagerström & Ågren, 1979)

$$\rho < \frac{1}{c} \left[ 1 - \frac{e^{a(1-c/2)} - 1}{a} \right] \qquad (2)$$

where $c$ is the relative variance in $a$ (i.e. $c = \sigma^2/a$). This condition can only be satisfied for $\rho \leqq +0.5$ (the limit $\rho = +0.5$ occurring for $a$ tending to zero): hence, there is a maximum degree of positive co-variation between the seed production of species 2 and that of species 1 consistent with survival of species 2.

The inequality (2) tells us that there is an upper limit on $\rho$ consistent with coexistence. To see what happens when

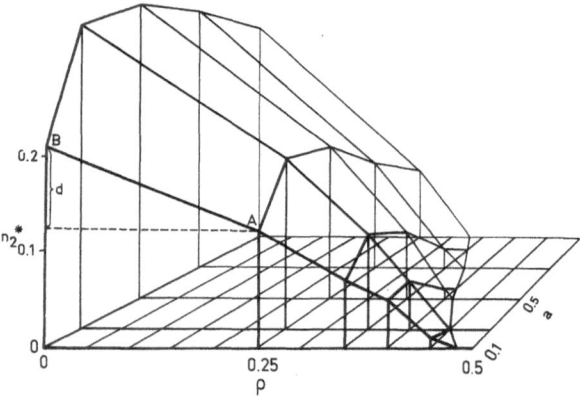

Fig. 1. The fraction of sites occupied by species 2 at steady state, as a function of $\rho$ and $a$, obtained from the solution to Equations (1) for $\lambda = 0.05$ and $c = 1$. Note that from the fig. the fitness change associated with an evolutionary adaptation can be derived also; e.g. if a specimen of species 2 evolves along the $\rho$-axis from the point A to B, the gain in fitness is given by the distance $d$.

84

$\rho$ is decreased below this limit, we have calculated from Eqs (1) the fraction of sites occupied by 2 at steady state as a function of $\rho$ and $a$. This is shown in Figure 1. Clearly, not only does survival of species 2 require $\rho < +0.5$, but also is further advantage conferred upon this species the further $\rho$ decreases below this limiting value. In contrast, increasing $a$ (as would result from relaxed competition for pollinators or pollination) will only be advantageous to species 2 for very small values of $a$, otherwise increasing $a$ will deteriorate the situation. These notions stress the intuitive point made previously, viz. that the marginal increase in fitness gained from producing an extra seed can be non-significant when competition for space between seedlings is intense. Therefore, only in some special cases should natural selection promote segregation of flowering times through the mechanisms dealt with by the two hypotheses discussed in the introduction.

## Discussion

Any evolutionary adaptation in species 2 leading to decreased $\rho$ is favourable. The question arises, how such adaptations could be interpreted in biological terms. We here consider pollination to be the step in the regeneration cycle that is limiting with respect to reproduction in the sense that less-than-average success in this step in a given year leads to a seed production below average and vice versa. It follows then that $c$ measures the relative variance (i.e. relative between-year variability) in pollination success, the major factor contributing to this variance usually being climatic variability, and we envisage $\rho$ to measure the degree of similarity between the two species with regard to phenology of flowering, such that equal phenologies implies $\rho = 1$ whereas phenological separation implies $\rho < 1$. Such an interpretation follows from the fact that the auto-correlation in most climatic variables declines so rapidly with time that already a few days of segregation of flowering times ensures that a significant degree of statistical independence in pollination success (small $\rho$) will result.

Thus, segregation in flowering times should be advantageous for an inferior competitor, leading to relaxed interspecific competition for colonization of empty sites by seeds. It should be noted that this advantage does not arise from any correlated segregation in time of seed dispersal or germination, nor does it involve the question of whether the inferior competitor flowers before, or after its superior competitor in a given year. The advantage is solely due to (1) that stochastic variations in pollination success exist, (2) that these variations at least in part can be attributed to weather variations, and (3) the autocorrelative properties of weather variations. As a consequence equal flowering times ($\rho = 1$) implies that a 'good year' for species 2 is always also a 'good year' for species 1 so that the reproductive effort by the former species is largely wasted. Similarly, a 'bad' year for species 1 is always a 'bad' year for species 2 so that the latter species cannot exploit the unusually high number of empty sites available to it. In contrast, separate flowering times ($\rho < 1$) implies that there is at least some chance for species 2 to have a 'good year' when species 1 has not.

Not unexpectedly, then, it can be shown that further advantage would be gained by species 2 if it evolved a reproductive strategy leading to negative $\rho$. Such a strategy would increase the chance for species 2 to have a 'good year' when species 1 has not; for $\rho = -1$ this would be a certain state of affairs. Negative $\rho$ would correspond to a situation where the two species have dissimilar physiologies, such that they respond to a given climatic variation in opposite directions (e.g. dry weather favours pollination in one species but disfavours it in the other) or, alternatively, that they respond to different climatic variables which are in themselves negatively correlated. Since we are here primarily concerned with ecologically similar species with presumably also similar physiologies, we do not further pursue the case of negative $\rho$ in this paper. Clearly, however, this hypothesis warrants attention, since it suggests an alternative (but possibly more expensive, sensu Roughgarden 1976) strategy for species 2.

It can be shown that the inequality (2) can only be satisfied if $c \neq 0$. That is, the potential for a segregated flowering time to be an adaptive strategy in species 2 vanishes if the sensitivity of the two species to environmental stochasticity is zero; stochasticity does not appear as 'noise' in an otherwise deterministic system, but as a fundamental property of the environment that opens new evolutionary routes for species 2 to employ in order to survive. Thus, we provide here an example that evolution may select for specialization to stochasticity so that the very non-predictability of an organism's environment can be taken advantage of (cf. the theories of 'fugitive equilibria' (Hutchinson 1961; Horn & MacArthur 1972, Levins & Culver 1971) 'r- and K-selection' (Gadgil & Solbrig 1972) and 'C-, S- and R-selection' (Grime 1977)).

To which of the observed instances of phenological spread at flowering is our hypothesis applicable? To see this, we recall that it is based on two assumptions. First, it assumes that pollination success varies randomly with time. That assumption is likely to hold generally for taxa pollinated by wind. In addition, we believe that it should also apply to many taxa pollinated by insects, since insects are indeed susceptible to short term weather changes (Johnson 1969) and hence our $\sigma$-parameters may as well be interpreted as measures of variations in insect activity. In the case of taxa pollinated by birds, we are aware of no data on activity variability in the relevant species; however, to the extent that these species exhibit such variability, our hypothesis should be relevant.

Second, our hypothesis assumes that there is competition for space between seedlings and young plants. Clearly, such competition should be intense in several major types of plant communities, e.g. mature forests and grasslands, but it is also clear that competition may be negligible in, e.g., deserts and other communities of arid regions. We are touching here the complex question of the degree of density-dependence in the regulation of population size, a topic that is beyond the scope of this paper. It should be noted, however, that while we have used an extreme example for the numerical·demonstration of our point (viz. that species 1 always wins, given that it is present) our hypothesis applies potentially also to cases where competition is less intense. In fact, we are dealing with a continuum of 'intensities of competition' and the theory developed in this paper potentially accounts for the whole of this spectrum, whereas the previous hypotheses are special cases accounting only for one extreme, viz. where competition is zero. To answer the question of exactly where on this continuum the fitness gained from producing extra seeds exceeds that from relaxing competition between seedlings will require the inclusion of other elements, such as cost-benefit analyses where the total reproductive·effort is constrained, e.g., by energy limitations.

In conclusion, we advance the general hypothesis that phenological spread at flowering, by virtue of the stochastic nature of short term weather variations, relaxes interspecific competition for available space to be colonized by seeds. This hypothesis should apply to species pollinated by wind, and insects, and possibly also to some of those pollinated by birds, and its applicability is independent of the details of the pollination biologies of the species concerned. It provides an explanation for the occurrence of phenological spread at flowering (1) in communities where there is competition for space between seedlings, and (2) in communities where neither the preva-lence of competition for pollinators nor of frequent interspecific pollinations seems plausible.

## Summary

Plant species co-inhabiting a given geographical region often have distinctly different times of flowering. It is shown that such phenological spread, due to short-term stochastic variation in weather variables, relaxes competition for empty sites to be colonized by diaspores. For sufficiently large spreads stable coexistence becomes possible. The applicability of the proposed hypothesis to the observed instances of phenological spread is discussed and shown to extend beyond that of other current theories.

## References

Fagerström, T. & G.I. Ågren, 1979. Theory for coexistence of species differing in regeneration properties. Oikos 33: 1–10.

Falinska, K. 1972. Fenologiczna reakcja gatunkow na zroznicowanie gradow. Phytocoenosis 1: 5–35.

Gadgil, M.D. & O.T. Solbrig, 1972. The concept of r- and K-selection: evidence from wild flowers and some theoretical considerations. Am. Nat. 106: 14–31.

Grime, J.P. 1977. Evidence for the existence of three primary strategies in plants and its relevance for ecological and evolutionary theory. Am. Nat. 111: 1169–1194.

Harper, J.L. 1977. 'Population biology of plants'. Academic Press, London-New York-San Francisco, 892 pp.

Horn, H.S. & R.H. MacArthur, 1972. Competition among fugitive species in a harlequin environment. Ecology 53, 749–752.

Hutchinson, G.E. 1961. The paradox of the plankton. Am. Nat. 95, 137–145.

Johnson, C.G. 1969. 'Migration and dispersal of insects by flight'. Methuen, London, . . . pp.

Levin, D.A. & W.W. Anderson, 1970. Competition for pollinators between simultaneously flowering species. Am. Nat. 104: 455–467.

Levins, R. & D. Culver, 1971. Regional coexistence of species and competition between rare species. Proc. Nat. Acad. Sci. 68: 1246–1248.

Mosquin, T. 1971. Competition for pollinators as a stimulus for the evolution of flowering time. Oikos 22: 398–402.

Pemadasa, M.A. & P.H. Lovell, 1974. Factors controlling the flowering time of some dune annuals. J. Ecol. 62: 869–880.

Roughgarden, J. 1976. Resource partitioning among competing species – a coevolutionary approach. Theor. Pop. Biol. 9: 388–424.

Stiles, F.G. 1975. Ecology, flowering phenology, and hummingbird pollination of some Costa Rican Heliconia species. Ecology 56: 285–301.

Waser, N.M. 1978. Interspecific pollen transfer and competition between co-occurring plant species. Oecologia (Berl.) 36: 223–236.

Wells, T.C.E. 1972. Ecological studies on calcareous grasslands. Monks Wood exp. Stn. Rep. 1969–1971: 44–46.

Accepted 31 October 1979

# AN EXPLORATORY ANALYSIS OF GRASSLAND DYNAMICS: AN EXAMPLE OF A LAWN SUCCESSION

M.P. AUSTIN*

Division of Land Use Research, Institute of Earth Resources, CSIRO, P.O. Box 1666, Canberra City, A.C.T. 2601, Australia

Keywords:
Grassland, Lawn, Numerical classification, Succession, Transition matrices

## Introduction

In recent years there has been a marked increase in interest succession. A sequence is evident from the pioneering criticism of Egler (1954) of the Clementsian (1916) concept of succession, via the critical review of Drury & Nisbet (1973) to the more recent work of Horn (1975, 1976) and Connell & Slatyer (1977). These authors conclude that classical 'relay floristics' where each group of species facilitates the invasion of successive groups may be uncommon and that, the 'initial floristic composition' of a site may be the major determinant of the seral sequence (Connel & Slatyer 1977). A major difficulty is the paucity of suitable data for examining these hypotheses.

Horn (1975, 1976) in his studies of forest succession made extensive use of transition matrices to demonstrate the processes and problems of classical succession concepts. Several workers have used this approach to succession (Stephens & Waggoner 1970; Waggoner & Stephens 1970; Godron & Lepart 1973; Henderson & Wilkins 1975; Enright & Ogden 1979) and its linear differential equation equivalents (Shugart et al. 1973). The obvious limitations of the transition matrices, the linear dependency of the current state on the immediately previous state of the system and constant transition probabilities have been discussed by Horn (1976) and van Hulst (1980). The method has potential however for examining alternative hypotheses concerning successional sequences and predicting community behaviour when other methods are not available (Enright & Ogden 1979). Studies have mostly been concerned with trees and forest succession.

* I thank T.H. Booth, H.H. Shugart and A.N. Gillison for comment on the manuscript, K. Mayo for programming assistance and M. Adomeit for data preparation.

Transition matrix methods have not apparently been applied to grasslands.

There are many studies of permanent grassland plots (Williams 1978; Bakker 1978), some with complex management treatments (Poissonet et al. 1978; Londo 1978). There is a need to evaluate the existing concepts and techniques if these permanent plot studies are to yield appropriate information about the future management of natural grassland communities (Debussche et al. 1977). This paper uses the results of ten years of observations on a lawn to examine the possible roles of detailed observations, numerical classification and transition matrices in succession studies. It will also be argued that further modifications of our conceptual ideas on succession are needed.

## Methods and data

The floristic composition of a lawn from near London, U.K. was observed with varying degrees of detail over a period of ten years beginning in 1957, approximately twelve years after its establishment. Local frequency data were collected from 39 1-yard (0.91 m) square contiguous quadrats determined from a six by six grid of subquadrats for all species. Grasses were not measured prior to 1960. The environment is uniform except for a marked gradient in shade parallel to the long axis of the lawn and a trampling effect along the western edge (see Austin (1977) for further details and a diagram of the site).

For the period 1960–1966, a series of detailed observations were made. An ordination study of these data is described in Austin (1977). The same data were grouped into sets of four contiguous subquadrats, that is nine new subquadrats per quadrat each one foot by one foot square (ca. 30 cm.). These data for the years 1960, 1961, 1963, 1965 and 1966 were then subjected to DIVINF (divisive information analysis, Lance & Williams 1968) presence / absence classification. The data matrix for this consisted of twenty-three species and 1755 quadrats.

The transition matrix study is based on the results of the numerical classification. The probability of quadrats remaining in the same 'community' (defined by the DIVINF classification, Appendix 1), or changing to another was determined for the two periods 1960–1961 and 1965–1966. The set of transition probabilities for all the communities recognized constitutes a transition matrix. By multiplying a vector of community abundances by a transition matrix, the seral behaviour of the lawn can be simulated, accepting the assumptions of the method. A particularly clear account is given by Enright & Ogden (1979). Further details of the use of transition matrices can be found in Horn (1975, 1976) and references quoted there. The transition matrices for the two periods contain some communities with very low numbers of observations and some rational adjustment has been made to the number and values of the entries in these columns (the actual transition matrices used are available from the author).

### Results

*Natural history observations*

The population dynamics as expressed by local frequency are presented for selected species summed for the entire lawn in Fig. 1. Certain species increase greatly (e.g. *Eurhynchyium praelongum*) and stabilize while others decline (e.g. *Ceratodon purpureus*); this appears to be an example of competitive displacement between two species of moss. *Festuca ovina* is an example of a species showing a stable population. *Trifolium repens* after an initial population collapse (1958–1959) shows an exponential population growth. *Sagina procumbens* has a period of increasing population followed by an oscillatory stabilization. These results do not show the spatial patterns which exist (cf. Austin 1977) nor can they indicate the complex spatio-temporal behaviour exhibited by some species.

*Trifolium repens.* Fig. 2a-e shows the invasive spread of white clover (*Trifolium repens*) after a catastrophic collapse in population due to drought in the early summer

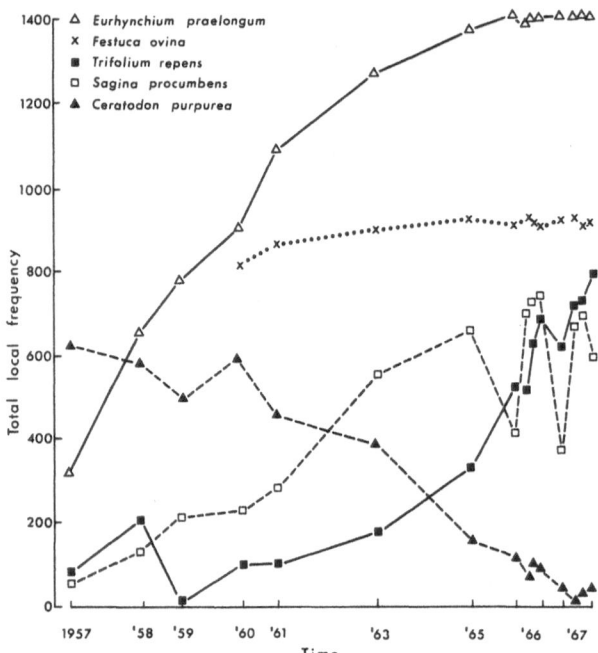

Fig. 1. Population dynamics of selected species in a lawn for the period 1957–1967.

of 1959. In July 1959 the population had fallen to 9 % of the September 1958 level (Fig. 2a). Three small patches of isolated stolons survived to July 1959. The mortality was related to the level of shading; the shaded patch (A in Fig. 2a) only suffered a 50 % reduction in area. A fourth patch undetected in 1959 is presumed to have survived as judged by later mapping (Fig. 2b).

From detailed observations on individual quadrats (Austin 1977), invasion by *Trifolium repens* leads to a marked reduction in many other species, particularly *Leontodon autumnalis* and other rosette lawn weeds. Under the mowing regime (approximately a ten day cycle), *T. repens* forms a dense canopy which apparently shades out the flatweeds. However, in 1965 after six years of occupancy of certain quadrats (Fig. 2c-e), *T. repens* began to decline in these areas. A fairy ring pattern developed with a dead centre, surrounded by a ring of growth with a mature canopy of leaves and an outer region of actively spreading stolons. The dead centre was re-invaded by some *Trifolium* stolons but a dense canopy was no longer developed. This phenomenon fits Watt's (1947a) classic description of the morphological pattern of *Pteridium aquilinum* (Watt 1947b) and other species (see also Austin 1968). It is compatible with Kershaw's (1959) interpretation of scales of pattern in *Trifolium repens*.

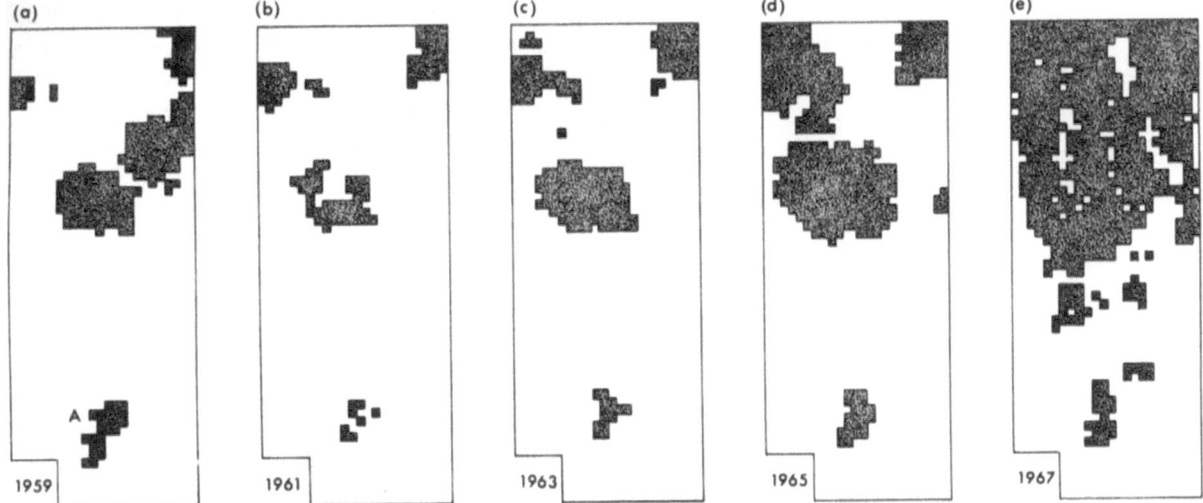

Fig. 2. Pattern of spread of *Trifolium repens* from 1958 to 1967 on a lawn.
(a) *Trifolium repens* present in September 1958 and surviving in July 1959 (dark shading) or absent (stipple). Shaded patch at A.
(b)-(e) Stages in the spread of *T. repens* from 1961 to 1967.

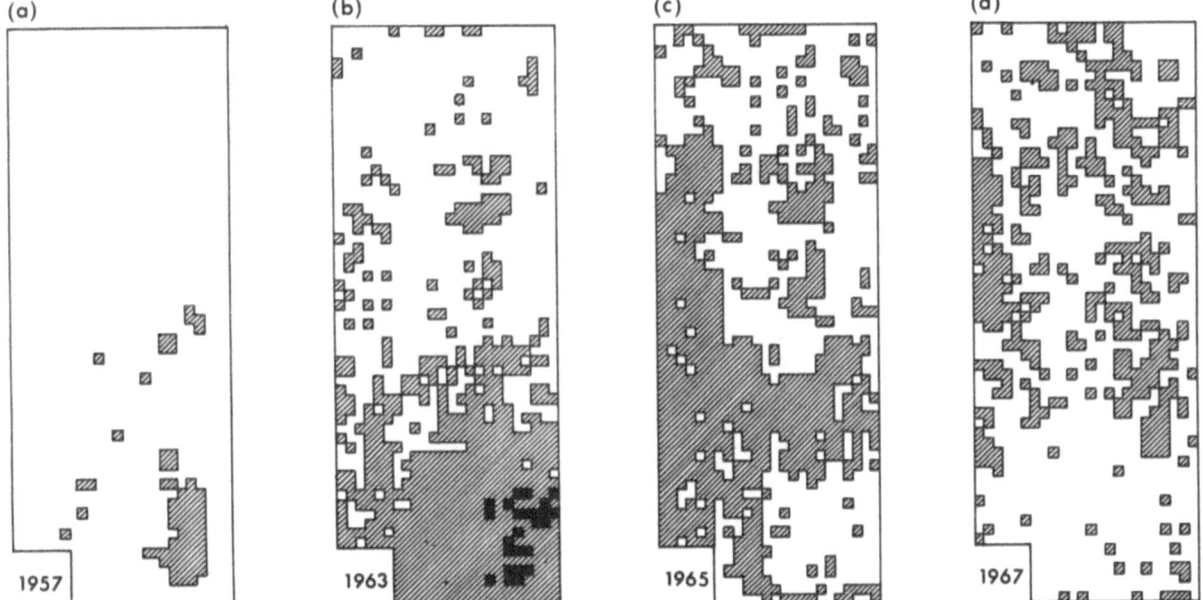

Fig. 3. Pattern of distribution of *Sagina procumbens* from 1957 to 1967 on a lawn. Areas where *Sagina procumbens* had died by May 1964 are shown by dark shading.

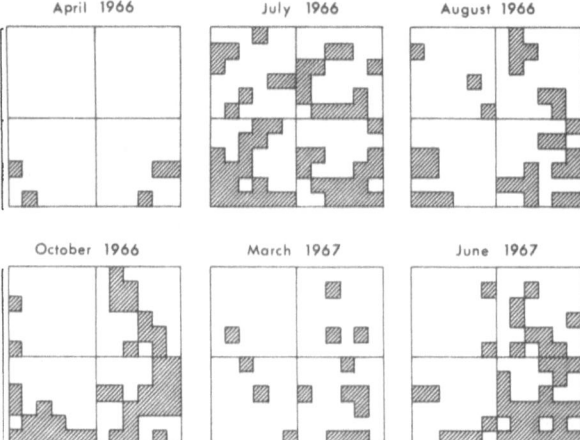

| April 1966 | July 1966 | August 1966 |
| October 1966 | March 1967 | June 1967 |

Fig. 4. Seedling establishment of *Sagina procumbens* (as indicated by local frequency) occurs in spring and late summer and deaths in winter and early summer (from lower right hand portion of Fig. 3).

This senescence phenomenon appears common in rhizomatous or stoloniferous plants (Watt 1947a; Kershaw 1973) and can produce a distinct non-random spatial mosaic of patches in a grassland vegetation (Austin 1968).

*Sagina procumbens*. This species showed a rapid population increase over the period 1957–1963, starting from an initial establishment in a highly shaded area of the lawn. It spread both vegetatively and by seed (Fig. 3a, b). In the highly shaded areas (lower portions of Fig. 3), under the frequent, close mowing regime, the species formed prominent patches of monospecific turf (apart from a few strands of the pleurocarpous mosses, *Eurhynchium praelongum* and *Brachythecium rutabulum*), eliminating other species, e.g. *Lolium perenne* and *Bellis perennis*. In the winter/spring period of 1964, the original area of turf was observed to be dying. A limited survey of the area in mid-May 1964 indicated that this was happening but that re-establishment by seed was also occurring. The local frequency measure was relatively insensitive to the biomass change in *Sagina procumbens*. The dense turves of *Sagina* did not develop again on the original area (Fig. 3a) in subsequent years (Fig. 4). The progressive spread of this wave of death can be documented from detailed mapping undertaken repeatedly in 1966 and 1967 (Fig. 4, Fig. 3b-d). Crops of seedlings were produced each year in spring and late summer, which suffered some mortality in early summer and major mortality during the winter period (Fig. 4, also Fig. 1). A major portion of the species

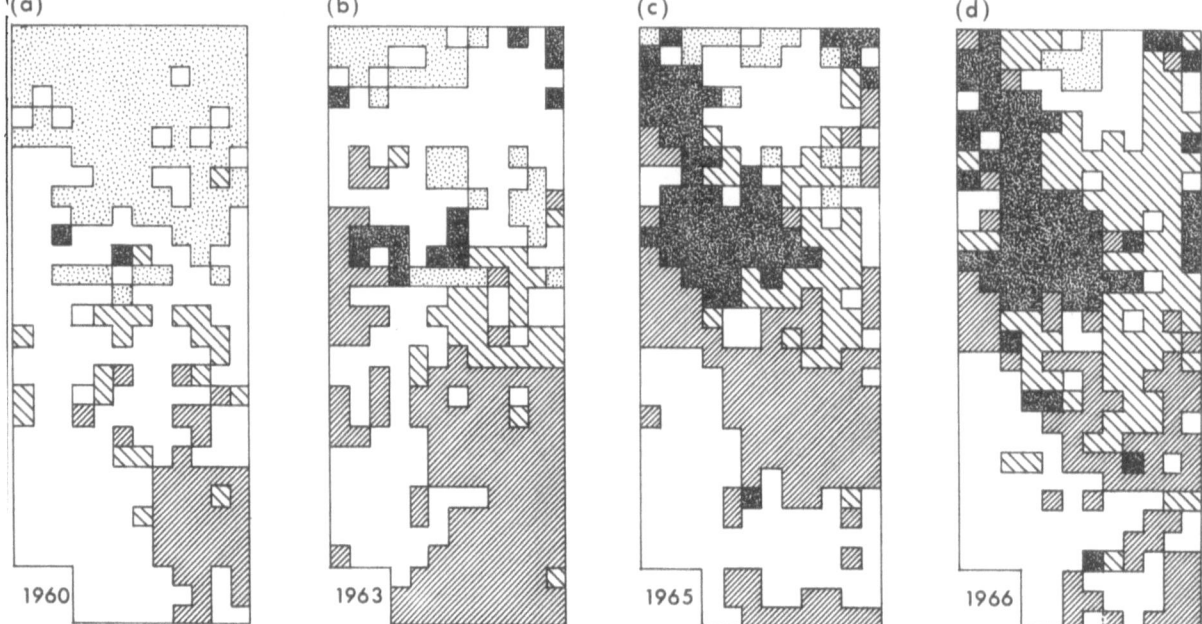

Fig. 5. Changes in the distribution of selected groups of DIVINF 'communities' over the period 1960 to 1966. Narrow diagonal hatching, *Sagina procumbens* defined communities (B, H, J in Appendix 1). Dense stipple, *Trifolium repens* defined communities (I and Q). Broad diagonal hatching, *Leontodon autumnalis* without *Ceratodon purpureus* communities (K, L). Light stipple, *Leontodon autumnalis* with *Ceratodon purpureus* communities (M, N).

90

Table 1. Lawn 'communities' defined by divisive information analysis (DIVINF) classification.

```
A - Leontodon autumnalis - Festuca ovina - Poa annua (perennial) - Sagina procumbens
B    "          "            "        "       "    "       "      + Sagina procumbens
C    "          "            "        "    + Poa annua (perennial) - (Pohlia nutans + Pottia truncata)
D    "          "            "        "       "    "       "      + (Pohlia nutans + Pottia truncata)
E    "          "          + Festuca ovina - Sagina procumbens - Ceratodon purpurea - Poa annua (perennial) - Trifolium repens
F    "          "            "        "       "       "          "    "       "      + Poa annua (perennial)
G    "          "            "        "       "       "          "    + Ceratodon purpurea
H    "          "            "        "    + Sagina procumbens - Cynosurus cristatus - Trifolium repens
I    "          "            "        "       "       "          "       "       "   + Trifolium repens
J    "          "            "        "       "       "          "    + Cynosurus cristatus
K + Leontodon autumnalis - Ceratodon purpurea - Brachythecium rutabulum
L    "          "            "        "          + Brachythecium rutabulum
M    "          "          + Ceratodon purpurea - Eurhynchium praelongum
N    "          "            "        "          + Eurhynchium praelongum - Cynosurus cristatus - Agrostis stolonifera
O    "          "            "        "            "        "       "          "       "         + Agrostis stolonifera
P    "          "            "        "            "        "       "          "    + Cynosurus cristatus
Q - Leontodon autumnalis + Festuca ovina - Sagina procumbens - Ceratodon purpurea - Poa annua (perennial) + Trifolium repens
```

+ species present.        - species absent.

distribution was affected by October 1967. A significant change had occurred in the population biology of the species. Some factor had converted a relatively long-lived perennial (8 years?) with low mortality of mature plants capable of forming a dense sward, into a short-lived perennial maintained by seed and with a low overwintering capacity. No evidence is available but it is possible that some pathogen may have established in 1963 and been responsible for the change.

The mapping of species distribution through time provides a description of the succession and combined with anecdotal evidence such as is provided for *T. repens* and *Sagina procumbens*, a descriptive explanation or natural history for a particular successional sequence. The description is on a species by species basis and for any even simple community the description rapidly becomes unmanageable.

## Multivariate classification

Multivariate methods offer a means of summarizing multispecies community dynamics (Austin 1977). A divisive information analysis classification was carried out (details not reported) and seventeen 'communities' of the one foot square subquadrats were recognized for the period 1960-1966. Many were transitory and sporadically distributed over the lawn. Fig. 5 (a-d) shows the dynamic behaviour of a few selected groups of the numerically defined 'communities'. Details of the communities and their defining species are given in Table 1. The gradual displacement of the *Leontodon autumnalis* plus *Ceratodon purpurea* communities is apparent. The spread and subsequent disintegration of the *Sagina procumbens* communities

can be seen while the progressive establishment of the *Trifolium repens* communities is well documented.

There is no possibility from such summary description of predicting the consequences of the observed changes. This form of summary description while not predictive in itself can be used in a predictive approach by utilizing the 'communities' in a Markov chain or transition-matrix approach. In grassland where individuals cannot be easily recognized, this is the only way to conveniently simulate and hence predict grassland dynamics.

## Transition matrix analysis

The observations on *Trifolium repens* (Fig. 2) and *Sagina procumbens* (Fig. 3 and 4) indicate that the behaviour characteristics of the plants were not the same for the periods 1960–61 and 1965–66. Their population dynamics were markedly different in the two periods. The transition matrices for the two periods are quite different. Fig. 6 shows the linkage diagrams corresponding to the two transition matrices. 'Absorbing states' (stable communities: those having no transitions to another community) can be recognized in the 1960–61 matrix, e.g. the two communities 'H' and 'Q'. The first is a *Sagina procumbens* defined community (Table 1) and the second a similar community defined by *Trifolium repens* and *Festuca ovina*. The final communities which would result from a simulation using the 1965–66 matrix are not readily identified.

A combined model, using the two matrices for the periods when the life history behaviour of the major species *Sagina procumbens* and *Trifolium repens* were different, provides an appropriate model. In matrix terms this is

91

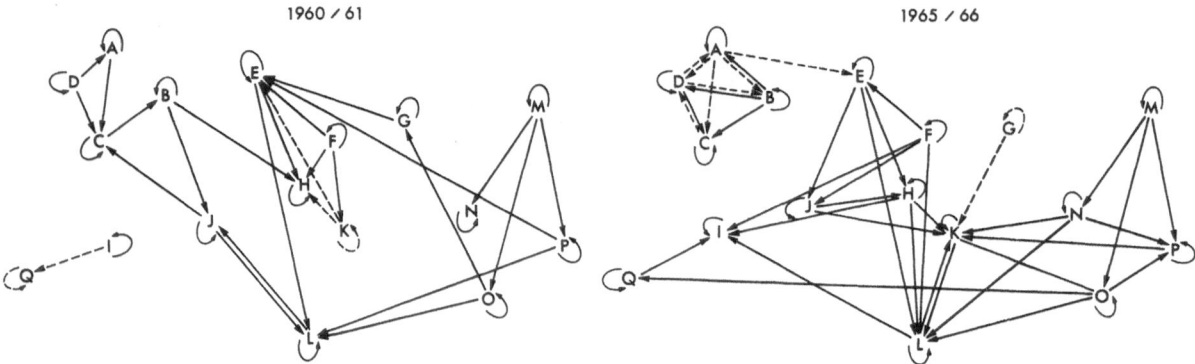

1960 / 61                                    1965 / 66

Fig. 6. Linkage diagrams for two transition probability matrices. Note links indicating persistence of particular state. Only major links ($\geq 0.1$) are shown. Estimated links are shown by dotted lines.

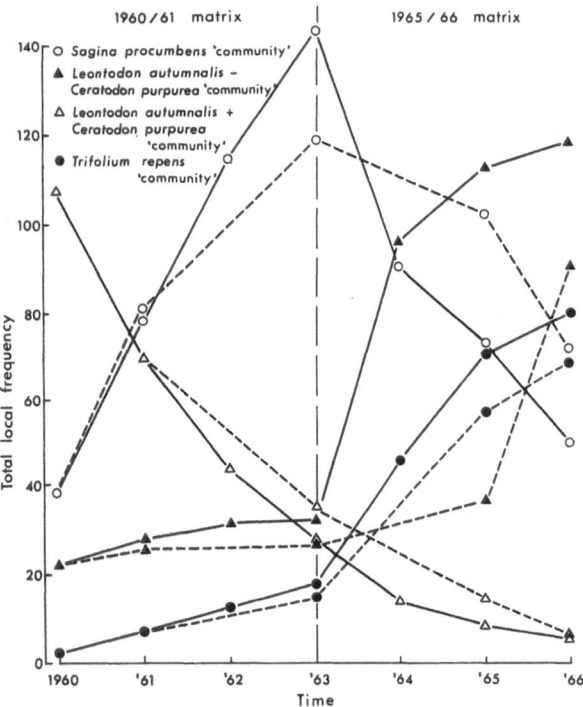

1960/61 matrix         1965/66 matrix

○ Sagina procumbens 'community'
▲ Leontodon autumnalis – Ceratodon purpurea 'community'
△ Leontodon autumnalis + Ceratodon purpurea 'community'
● Trifolium repens 'community'

Fig. 7. Simulation of the seral changes in a lawn from 1960 to 1966. Solid line shows simulation results, dashed line actual observations.

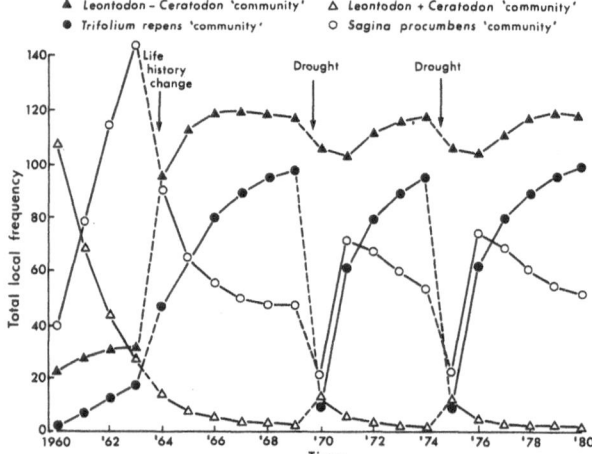

▲ Leontodon – Ceratodon 'community'    △ Leontodon + Ceratodon 'community'
● Trifolium repens 'community'         ○ Sagina procumbens 'community'

Fig. 8. Simulation of the possible future seral development of a lawn with the random inclusion of a drought catastrophe transition matrix.

**BBBBAAA.X$_0$** where **A:** transition matrix 1960–61
　　　　　　　　　**B:** transition matrix 1965–66
　　　　　　　　　**X$_0$:** community data for 1960.

The model can be compared with actual observations (Fig. 7). The **A** matrix provides a reasonable simulation of the community dynamics over the first three years. Use of matrix **B** produces qualitative agreement with actual observations (Fig. 7). There is however a distinct

lag in the actual observed changes as compared with the simulated observations. This can be attributed to the fact that in the transition matrix model, every subquadrat with *Sagina procumbens* has an equal chance to become converted to a different 'community' due to the death of that species. As indicated by Fig. 3, the (presumed) pathogen had a highly contagious distribution and took four years to spread over the lawn. The transition probabilities have a spatial and temporal dependence on the initial patterns of establishment.

From observations (Fig. 2a) however we can conclude that this model will not be satisfactory even as a linear model, for droughts are known to occur, e.g. as in 1959, which will drastically alter the abundance of species. Such droughts occur about once in seven years. A simulation using the **B** matrix (normal climatic conditions) and a **C** matrix representing the effect of summer drought would provide a better prediction of future (post-1966) community dynamics in the lawn. Though the drought transition matrix for the communities can not be estimated (no grasses were recorded), the observations on the drought of 1959 provides some subjective information on which to construct such a matrix. The effect of including such a matrix is shown in Fig. 8. The simulation suggests that there is a rapid return to near-equilibrium conditions from the catastrophic drought. Survival is dependent on the degree of shading (Fig. 2), so Fig. 8 provides a biased prediction of future dynamics.

**Discussion**

Given the requirements of grassland management, some predictive capacity is required of analysis methods. Current descriptive methods do not have this in any explicit form. Using a transition matrix to describe succession in grassland communities can provide estimates of future behaviour in a way which current descriptive analysis of permanent plots cannot. However, an important feature which requires further study is the data requirements for estimating the transition probabilities with sufficient accuracy. Given the limitations of the transition approach, careful, detailed observations and as much natural history information as possible are necessary to complement any transition matrix simulation study.

Forest succession studies (Horn 1976, Shugart et al. 1973) have used cover or individuals and assumed replacement by seedlings and saplings observed to occur under the canopy of mature trees. Recognition of individuals is not a practical possibility for grassland studies. Community types offer a means of analysing grassland without this problem. Use of an explicit technique (DIVINF classification) avoids some of the problems of what subjectively defined cover types actually mean in terms of composition, but more suitable classification methods may be found.

The data and simulations using transition matrices reflecting different climatic conditions are practical examples of some of the theoretical possibilities raised by Horn (1975, 1976). The real-world complexities of grassland limit the possible use of transition matrices for prediction. The spatial patterns which develop in grasslands (Watt 1947a, b, Kershaw 1973) and determine the successional behaviour of the sward (Figs. 2, 3, 5), are dependent on environment and initial patterns of establishment (Austin 1977). The detailed prediction of within-site population dynamics has not been attempted as it would not be possible without taking account of adjacency (current state of neighbours) when determining the probability of a change of state of a subquadrat. Development of methods incorporating spatial dependency appears to be the next methodological advance required in predictive succession studies.

Most discussions of plant succession (Horn 1976, Connell & Slatyer 1977; Drury & Nisbit 1973) while mentioning disease and senescence do not allocate a major role to these factors in determining the dynamics of a sere. When a species initially invades a new area and undergoes a population explosion, presumably it has been able to escape from control by its pathogens and predators. One may then expect a population collapse when either pathogens or predators reach the site. This may have been the case with *Sagina procumbens* in the lawn. The rise and fall of populations during a succession may reflect such epidemic phenomena far more frequently than they reflect 'facilitation' or a 'competitive hierarchy'. Succession involves more than a single trophic level, a feature emphasized in recent discussions but not often documented (Connell & Slatyer 1977). The role of grazers or pathogens in determining certain successions is well-recognized, but they must be expected to play a role in all vegetation dynamics, and it may be necessary to consider that a species' predators, parasites and pathogens will facilitate its replacement by later successional species. The absence of grazers and pathogens may retard the the rate of succession, and allow development of unusual communities with unexpected dominants, e.g. *Sagina procumbens*. The senescence phenomenon observed with *T. repens* may also be caused by disease but the reinvasion though not with the same vigour as during the building and mature phases of the patch (Watt 1947a) suggests a need for a more complicated explanation. The importance of the phenomenon will depend on the scale at which prediction is needed. If prediction is required at a scale similar to the 'fairy-ring' patch then more complex models will be required.

Experiments on permanent plots for the purpose of managing natural or semi-natural communities require careful design for them to yield maximum information. The experiment reported here has some of the required features, e.g. detailed observations (though not sufficiently regularly recorded), contiguous quadrats for recognition of spatial phenomena and a range of environments to enable some generalization of the conclusions.

However, lack of replication of sites may lead to capitalizing on chance or the special features of a single site. The transition matrix approach does not provide the most suitable method for making *spatio-temporal* predictions of vegetation patterns during a grassland succession and would not be reasonable without an adequate natural history knowledge of the vegetation.

The observations presented here together with the examples of transition matrix analysis emphasize that suitable analysis methods *and* detailed observations are necessary, if meaningful predictions about particular successions are to be made. In the absence of new methods, the judicious use of transition matrices with appropriate community types appears to be a useful method for analysis of permanent grassland plots.

## Summary

The floristic composition of a lawn is studied by means of permanent plots over ten years. Observations show that changes in the life history behaviour of certain species (*Trifolium repens*, and *Sagina procumbens*) have important effects on the vegetation dynamics of the lawn. These changes are ascribed to senescence and possibly a pathogen. Numerical classification (divisive information analysis) is used to define suitable classes for transition matrix simulation of grassland succession. The influence of life history changes and catastrophic drought is incorporated by estimating appropriate transition matrices. The limitations of the approach due to spatial patterns are emphasized.

## References

Austin, M.P. 1968. Pattern in a Zerna erecta dominated community. J. Ecol. 56: 739–757.

Austin, M.P. 1977. Use of ordination and other multivariate descriptive methods to study succession. Vegetatio 35: 165–175.

Bakker, J.P. 1978. Changes in a salt-marsh vegetation as a result of grazing and mowing. A five-year study of permanent plots. Vegetatio 38: 77–88.

Clements, F.E. 1916 Plant succession. Carnegie Inst. Washington, Publ. 242, 512. pp.

Connell, J.H. & R.O. Slatyer. 1977. Mechanisms of succession in natural communities and their role in community stability and organisation. Amer. Nat. 111: 1119–1144.

Debussche, M., M. Godron, J. Lepart & F. Romane. 1977. An account of the use of a transition matrix. Agro-Ecosystems 3: 81–92.

Drury, W.H. & I.C.T. Nisbet. 1973. Succession. J. Arnold Arboretum 54: 331–368.

Godron, M. & J. Lepart 1973. Sur la representation de la dynamique de la végétation au moyen de matrices de succession. In R. Tuxen (ed.) Sukzessionsforschung. Ber. symp. Int. Ver. Vegetationskunde, Rinteln, 1973, p. 269–287, Cramer, Vaduz.

Egler, F.E. 1954. Vegetation science concepts. 1. Initial floristic composition- a factor in old-field vegetation development. Vegetatio 4: 412–417.

Enright, N. & J. Ogden 1979. Applications of transition matrix models in forest dynamics: Araucaria in Papua New Guinea and Nothofagus in New Zealand. Aust. J. Ecol. 4: 3–23.

Henderson, W. & C.W. Wilkins 1975. The interaction of bushfires and vegetation. Search 6: 130–3.

Horn, H.S. 1976. Markovian properties of forest succession. In: M. Cody and J. Diamond (eds.), Ecology and evolution of communities, p. 196–211. Harvard University Press, Cambridge.

Horn, H.S. 1976. Succession. In: R.M. May (ed.), Theoretical ecology: Principles and applications, p. 187–204. Blackwell Scientific Publishers, Oxford.

Kershaw, K.A. 1959. An investigation of the structure of a grassland community. II. The pattern of Dactylis glomerata Lolium perenne and Trifolium repens. III. Discussion and conclusions. J. Ecol. 47: 31–53.

Kershaw, K.A. 1973. Quantitative and dynamic ecology. 2nd. ed. Edward Arnold, London, 308 pp.

Lance, G.N. & W.T. William 1968. Note on a new information-statistic classificatory program. Comput. J. 11: 195.

Londo, G. 1978. Möglichkeiten zur Anwendung von vegetations-kundlichen Untersuchungen auf Dauerflächen. Vegetatio 38: 185–190.

Poissonet, P., F. Romane, M. Thiault & L. Trabaud 1978. Evolution d'une garrigue de Quercus coccifera L. soumise à divers traitements: quelques resultats des cinq premières anneés. Vegetatio 38: 135–142.

Shugart, H.H. Jr., T.R. Crow & J.M. Hett 1973. Forest succession models a rationale and methodology for modelling forest succession over large regions. Forest Science 19: 203–212.

Stephens, G.R. & P.E. Waggoner 1970. The forests anticipated from 40 years of natural transitions in mixed hardwoods. Bull. of the Connecticut Agricult. Exp. Stat., New Haven. 58 pp.

Van Hulst, R. 1980 Vegetation dynamics or ecosystems dynamics: Dynamics sufficiency in succession theory. In: E. van der Maarel (ed.) Advances in vegetation science: Succession. Vegetatio 43: 95–102.

Waggoner, P.E. & G.R. Stephens 1970. Transition probabilities for a forest. Nature 225: 1160–1161

Watt, A.S. 1947a. Pattern and process in the plant community. J. Ecol. 35: 1–22.

Watt, A.S. 1947b. Contributions to the ecology of bracken. IV. The structure of a community. New Phytologist 46: 97–121.

Williams, E.D. 1978. Botanical composition of the park grass plots at Rothampsted 1856–1976. Rothampsted Expt. Stat., Harpenden, 61 pp.

Accepted 6 November 1979

# VEGETATION DEVELOPMENT IN A FORMER ORCHARD UNDER DIFFERENT TREATMENTS: A PRELIMINARY REPORT*

Eddy VAN DER MAAREL**

Division of Geobotany, University of Nijmegen, Toernooiveld 1, 6525 ED Nijmegen, The Netherlands

Keywords:

*Aegopodion podagrariae, Arrhenatherion elatioris*, Development, Dynamics, Experiment, Fertilization, Mowing

## Introduction

An experiment on the vegetation development of the tall herb understory of an orchard was started in 1969. In that year the University arranged with the State Forestry Service that an apple orchard belonging to the woodland nature reserve Wylerberg was cut and a series of experimental plots was established. The idea was inspired with the famous Park Grass experiments at Rothamsted (Williams 1978, Thurston, Williams & Johnston 1976) and a simplified version for didactical purposes set up by C.D. Pigott in the Botanical Garden of the University of Lancaster. Because we could not find a suitable grassland on poor soil we decided to start with a very rich soil and try to induce several changes leading to poorer soil conditions.

The orchard area is situated at the north side of the push moraine east of Nijmegen, The Netherlands, at 40 m above sea level. The apple orchard in which the experiments were arranged had been heavily manured for years until 1968. The top soil layer of the eastern part consists of loamy sand which is relatively moist because of the run-off of

* Nomenclature follows Heukels-van Ooststroom, Flora van Nederland, 18e druk, Wolters-Noordhoff, Groningen, 1975.
** Vegetation and soil analyses were mainly carried out by students in the framework of their study for the degree of doctorandus in biology: Ineke Opbroek, Ferdinand Baggerman (1970), Carla Veldhuis (1971-1972), the late Doeke Dijken (1971-1973), Jos Rijpert (1972), the Rev. Piet van der Aart, Jo Louppen (1973), Frank Verhoeven (1974), Maurits Heine (1975), Ad Peters (1976), Huub Muyres, Louis Fliervoet (1976, 1977), Emilia Peters (1977), Niek Joanknecht, Thijs Lavrijsen (1977, 1978).
The present contribution is largely based on their internal reports. I also acknowledge the help of drs. A.J. Kempers (Div. of Soil Science) and Dr. G.W.M. Barendse (Botanical Garden).

water from higher parts (up to 80 m) of the moraine. The top (8 cm) contains 3–5 % organic material, the pH is between 6 and 7. Values for N and P total are in the order of 0.2 and 0.1 % respectively. The western part of the experimental area consists of coarse sand and it is better drained. This part was already free of trees by the time the experiments started. The vegetation in the orchard consisted mainly of ruderals such as *Urtica dioica, Lamium album* and *Lamium maculatum* and tall grasses such as *Dactylis glomerata* and *Elytrigia repens*. They characterize socalled nitrophilous edge communities of the alliance *Aegopodion podagrariae*. In the area without fruit trees an *Arrhenetherum* grassland occurred.

The potential vegetation of the area is a woodland of the *Fago-Quercetum* type and on places where water runs off *Circaeo-Alnion* (Westhoff & den Held 1969). Most characteristic species occur in the immediate surroundings of the experimental area, e.g. *Lamium galeobdolon, Stellaria holostea, Pteridium aquilinum, Senecio fuchsii, Equisetum telmateia, Phyteuma nigrum*.

## Set up of the experiment

In 1970 12 plots of 10 m × 7 m were laid down in the orchard part and in 1972 8 plots of the same size were laid down in the grassland part. Table 1 summarizes the treatments these plots were subjected to. The standard treatment is yearly mowing at a time shortly after optimal development of the dominant species (July-August). The phytosociological perspective of this type of treatment is the development of the alliance *Arrhenatherion elatioris*. It is supposed that through removal of the bulk of the biomass produced in one season the supply of available nutrients is

Table 1. Survey of treatments of experimental plots.

| Plot | Mowing | Years | Treatment | Years |
|------|--------|-------|-----------|-------|
| 1 | S[1] | 70- | | |
| 2 | -[2] | | | |
| 3 | S | 70- | | |
| 4 | S | 70- | marl[5] | 71-75 |
| 5 | S | 70- | peat[6] | 71-75 |
| 7 | S | | | |
| 8 | S | 72- | Sod cut | 71 |
| 9 | (S)[3] | 70-72 | furrowed | 73 |
| 10 | S | 70- | NPK[7] | 71-73 |
| 11 | S | 70- | | |
| 12 | S | 70-- | Ca,Mg[8] | 71-75 |
| 13 | (S) | 70-72 | | |
| 14 | (S) | 70-72 | urea[9] | 73-75 |
| 15 | S | 70- | | |
| 16 | S | 70-(not 73) | urea | 73-74 |
| 17 | S | 70- | | |
| 18 | S | 70- | urea | 73-74 |
| 19 | (S) | 70-72 | | |
| 20 | S | 70- | | |
| PB | 3x[4] | 70- | | |

[1.] mown in summer; [2.] not mown since 1970; [3.] mown in summer 1970-1972, not mown since 1973; [4.] PB: plots between large plots: mown 3-4 x/yr; [5.] 2 kg/m$^2$/yr; [6.] 1,5 l/m$^2$/yr; [7.] ASF pellets 0,15 kg/m$^2$/yr; [8.] Ca+Mg carbonate 0,7 kg/m$^2$/yr; [9.] 0,15 kg/m$^2$/yr.

gradually but steadily decreased. In some plots we started with mowing two times a year, i.e. in May and August, but because we did not find differences with plots which were already mown in August we decided to finish this double mowing in 1974. Finally some small plots were laid down between the large plots to follow the effect of mowing 3-4 times a year, the mowing scheme deviced for the pathes and interspaces. Because of management difficulties this scheme could not be maintained in the years 1976-1978.

The second main 'treatment' is the addition of fertilizers. Plot 10 was fertilized once a year in April with ASF pellets, containing N, P and K (as $NH_4 + NO_3$, $PO_4$ and $K_2SO_4$, 6 : 6 : 10 : 18). To study the possible effect on ruderals of a pure N fertilization urea was used on some plots which did not contain typical ruderal species. After two or three successive treatments the expected changes had become clear and fertilization was stopped.

To promote nitrogen mineralization in one plot the top 10 cm of soil was furrowed and in two other plots lime was added. To change the nitrogen/carbon ratio peat was added to the soil of plot 5.

To speed up oligotrophication, in plot 8 the top 10 cm of soil were removed.

In various plots no mowing was applied any longer, mainly in order to observe under which circumstances a development towards the *Aegopodion podagrariae* would take place.

In plot 6 a combination of treatments was applied on a smaller scale. This plot will be left out of consideration.

## Observations

Each plot was divided into four quarters. A Braun-Blanquet analysis (Westhoff & van der Maarel 1978) of each subplot was made yearly. The spring aspect with *Lamium album* and *L. maculatum* was analysed separately. Spring data were integrated with the data of the main analysis in July-August. Both cover/abundance and sociability were recorded with refined scales. In the numerical analysis cover-abundance data in their ordinal transform values were used: values 1–4 correspond with abundance classes r, +, 1 and 2 m; values 5–9 correspond with cover classes 5–12,5 %, 12,5–25 %, 25–50 %, 50–70 % and 75–100 % (Van der Maarel 1979). Soil and crop chemical analyses were done irregularly and will not be treated in detail here.

## Data processing and -analysis

Data on species occurrence in all subplots were first screened for inconsistencies before being stored on punched cards and magnetic tape and a 'definitive list of recognisable taxa' was composed. This list does not include doubtful identifications, seedlings (the establishment of which will be studied separately) and mosses (of which only a few species occur). By 1979 ca. 110 vascular non-woody species were recorded.

For each subplot the distribution in time of all species was studied, species richness determined and a syntaxonomical spectrum calculated. (The latter reflects the quantative relations between the various syntaxonomical groups, such as *Aegopodion* and *Arrhenatherion* to which the species can be assigned, see e.g. van der Maarel 1978). Subplot data of each year were subjected to a TABORD-classification and a PCA (ORDINA) ordination (see van der Maarel, Janssen & Louppen 1978).

One of the results of the latter treatment was that the four subplots of most of the plots appear to be very similar so that one subplot can be taken as representative.

The main treatments to be reported on are a classification and an ordination of the A-subplots from the years 1970, 1972, 1974, 1976 and 1978.

Table 2. TABORD-classification of A-subplots of experimental plots 1970–1978 (see text).

| Cluster numbers | 1 | 2 | 3a | 3b | 4a | 4b | 5 | 6a | 6b | 6c | 7 | 8 | 9 | 10 |
|---|---|---|---|---|---|---|---|---|---|---|---|---|---|---|
| Represented plots | 2 | 1 | 12 | 14 | 9 | 16 | 13 | 1 | 15 | 10 | div | 17 | 20 | 8 |
| | | 11 | | | | 18 | | 12 | 27 | | | 19 | | |
| Years | 0-8 | 0-2 | 0-6 | 4-8 | 2-8 | 4-6 4-8 | 2-8 | 4-8 | 4-8 | 4-8 | 6 | 2 | 2-8 | 2-8 |
| Number of relevés | 5 | 12 | 6 | 3 | 4 | 5 | 5 | 23 | 6 | 3 | 4 | 3 | 4 | 4 |
| Mowing regime | - | + | + | - | - | + | - | + | + | + | + | + | + | + |
| | | | | | | | | | | | | | | |
| Cuscuta europaea | 3-6 | | | | | | | | | | | | | |
| Calystegia sepium | 3-4 | | 2 | 2 | 3-6 | 3-6 | 3-5 | 2-3 | 2 | | | | 2 | |
| Lamium maculatum | 5-9 | 3-6 | 0-3 | 3-6 | 5-/ | | 2-6 | 1-3 | | | | | | |
| Urtica dioica | 5 | 5-8 | 5 | 8 | 6 | 7-9 | 3-7 | 3-5 | 3-5 | 5-3 | 4 | 3 | 1 | |
| Heracleum sphondylium | 5 | 2-5 | 2-7 | 2-7 | 5 | 3-7 | 5-6 | 2-6 | 3-6 | 5 | 1-3 | 7 | 3-5 | 1-3 |
| Dactylis glomerata | 5-7 | 3-5 | 0-5 | 3-5 | 5-7 | 3-5 | 4 | 5-7 | 4-6 | 4-6 | 4-6 | 5 | 2 | 3-8 |
| Poa trivialis | 4 | 3-5 | 3-7 | 4-7 | 4 | 4 | 4 | 4-6 | 3-6 | 4-6 | - | 3 | 2-4 | 2-4 |
| Elytrigia repens | 5-7 | 4 | 7-9 | 6-2 | 4 | 4 | 4 | 3-6 | 3-5 | 4-7 | 3-5 | 5 | 0-4 | |
| Holcus lanatus | 2 | 3-6 | 3-5 | 3-7 | 0-3 | 2 | 5-8 | 5-7 | 4-7 | 3-7 | 0-3 | 5-7 | 4 | 4-8 |
| Lamium album | | 2-4 | | | 5 | 0-2 | 2-4 | 2-4 | 3 | | | | | |
| Taraxacum officinale | | 2-5 | 2 | 1-2 | | 0-1 | 2 | 2-5 | 2-3 | 3-5 | 2 | 3 | 2 | 5 |
| Arrhenatherum elatius | | | 0-6 | 3-6 | 4-7 | 6-8 | 4-6 | 3-7 | 7-8 | 7 | 7-9 | 3-7 | 9 | 3-6 |
| Artemisia vulgaris | | | | | | 5-9 | | | | | | | | |
| Tanacetum vulgare | | | | | | 1-2 | | | | | | | | |
| Achillea millefolium | | | | | | 1-2 | | | 2 | | 1-2 | 3-5 | 2-3 | |
| Cerastium holosteoides | | | | | | | | 1-3 | 2-4 | 1-3 | | | | 2-3 |
| Agrostis tenuis | | | 0-2 | 2-3 | 3 | | | 2-4 | 0-4 | | 1 | | 4 | 2 |
| Vivia hirsuta | | | | | | | | 2-5 | 0-2 | 2 | | 5 | 4-5 | 3-6 |
| Campanula rapunculus | | | 2 | | | | | | | | | 2 | 1-4 | |
| Vicia sativa-angustifolia | | | | | | | | | | | | | 2 | 0-2 |
| Melandrium rubrum | | 0-3 | | | | | | | | 5 | | | | 3 |
| Ranunculus repens | | | | | | | | | 2-4 | | | | | 3-5 |
| Plantago maior | | | | | | | | | | | | | | 1-3 |
| Linaria vulgaris | | | | | | | | | | 4 | | | | 1-4 |
| Trifolium dubium | | | | | | | | | | | | | | 3-5 |

## Results

### Classification

Table 2 presents a summary of a TABORD-classification, with the similarity ratio as a resemblance function and a fusion limit of 0.60. Ten end-clusters resulted. They are indicated with the number(s) of the plots included in the clusters over the years 0 : 1970, 2 : 1972, etc. In three cases a substructure is presented which leads to a survey of 14 types. The occurrence of species in each type is indicated as follows: if the frequency exceeds 79 % the transformed cover-abundance, or its range, is given. In cases of a frequency of 50-70 % the indicated range has 0 as its lower value. Other occurrences are neglected.

Type 1 includes plot 2 throughout the years and it shows the typical development of the non-mown vegetation. Ruderals and climbing species predominate. The type is distinguished from all others by *Cuscuta europaea*. Type 2 includes most of the plots of the 1-12 series, but only in the early years, when the influence of the heavy manuring was still overriding. This influence is reflected by the combination of *Lamium maculatum* and *Urtica dioica* in high quantities. Type 3a is mainly built up by 4 relevés of plot 12. This plot is related to most plots of type 2, but differs in the presence of *Arrhenatherum*, which occurred in the western half of the area already by 1970. Type 3b includes the three relevés of plot 14 after (in 1973) urea was applied here. Although this treatment was stopped in 1975, its influence is still there, as follows from the high values for *Urtica* and *Lamium maculatum*. This species was completely absent in plot 14 in the years 1970-1972 but entered the plot already in the summer of 1973.

Cluster 4 combines two types which are different in treatment. Type 4a contains the 1972-1978 relevés of plot 9, which was furrowed only in 1973 and still mown in 1972. Type 4b includes relevés of urea treated plots 16 and 18, which are mown. It would correspond with type 2 rather than with type 4a. Apparently the combination of *Arrthenatherum* with dominant *Urtica* and *Calystegia sepium* determines its position.

Cluster 5 contains all relevés of plot 13, which belongs to the western half (*Arrthenatherum*!) but has ruderals such as *Lamium maculatum* as well as species of relatively dry ruderal places; (association *Tanaceto-Artemisietum*).

Cluster 6, the largest one, combines mown plots from the eastern half in the later years (types 6a, c) with typical rough grassland plots of the western half (type 6b). Type 6c, NPK treated plot 10 in the later years, contains some *Melandrium rubrum*, *Ranunculus repens* and *Linaria vul-*

*garis*, species which are known from different habitats, but also occur together (since 1974) in the neighbouring plot 8.

Type 7 includes relevés from 4 different plots, but 3 of them are from 1976. Its floristic differentiation is unclear.

Type 8 contains early relevés of the edge of the eastern half in which *Campanula rapunculus* occurs. Type 9, referring to plot 20, both early and later years, is rather similar. Type 10, containing relevés of plot 8, in which some species typical of open grasslands, e.g. *Plantago major*, and a high amount of *Fabaceae* occur.

### Ordination

The dynamical relations of the various plots are shown in Fig. 1, which presents the position of relevés (most of the subplots in years 1970-1978) along the first two components. To facilitate interpretation data for some plots strongly resembling others are left out, while the group of mown grassland or grassland-becoming plots is presented separated from the group of not-mown plots (Fig. 1a, 1c). Fig. 1b helps in the interpretation of the model by showing isolines for some important species and marking the area in which the alliance character-species of the *Arrhenatherion* are concentrated.

The following main trends can be noticed:
1. Plots 2 and 20, forming the main floristic contrast within the whole series, being under a stable management, are floristically stable.
2. Plots of type 2, of the eastern half, being subjected to mowing, develop in the direction of plot 20, the extreme plot in the western half. The closer the position of the plot to the western half the further the development proceeds, with plot 12 almost reaching the situation of plot 20 in 1978.
3. The different treatments of type 2 plots, fertilization + mowing, liming + mowing, only mowing, do not seem to lead to different developments, mowing being the overriding factor.
4. The removal of the top soil (plot 8) and subsequent mowing leads to an *Arrhenatherion* grassland much more rapidly than just mowing does.
5. Not mowing equally leads to a convergence, but here the original differences between the plots are maintained longer.
6. Heavy fertilization immediately provokes the establishment of the ruderals *Urtica* en *Lamium maculatum*. As shown by the course of plot 16 mowing after stopping fertilization brings a plot rapidly back to the *Arrhenatherion* type grassland.

98

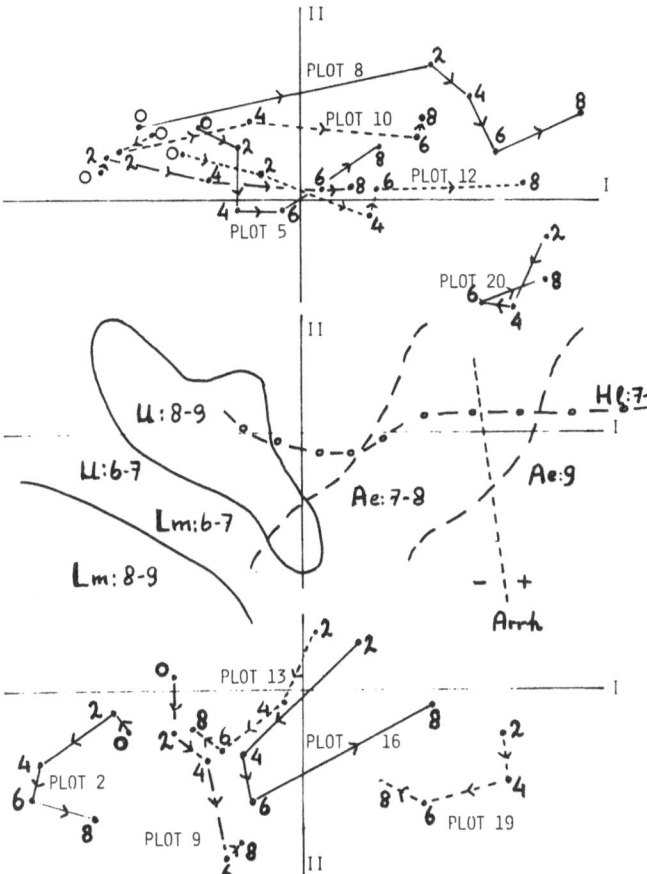

Fig. 1. Ordination of A-subplots of experimental plots 1970-1978, principal components 1 and 2. Top: position of plots 1, 5, 8, 10, 12 and 20; bottom: position of plots 2, 9, 13, 16 and 19; middle: isolines for Lamium maculatum (Lm), Urtica dioica (U), Arrhenatherum elatius (Ae) and Holcus lanatus (H1), and line separating areas with and without representation of character species (other than Arrhenatherum) of the alliance Arrhenatherion (Arrh.).

*Behaviour of species*

These general trends can be specified and better understood by following individual species populations in contrasting plots. Table 3 presents the dynamics of four important species in six representative plots. The following observations can be made.

1. Plot 1, representative of the eastern mown plots shows the main floristic trend, i.e. *Urtica dioica* rapidly decreasing, but not disappearing, and *Arrhenatherum* appearing and steadily increasing.

2. The effect of not-mowing as in plot 2 is clearly the persistance of *Urtica* and an optimum of *Lamium*

*maculatum* shortly after the non-mowing regime was introduced. Plot 13 shows how both species can first rapidly increase if they are present already but in low quantities. (From plot 19 we know that if such species are not present they do not easily enter after mowing has been stopped).

3. Plot 16 shows how *Urtica* can drastically increase after the addition of urea, but decrease equally drastically after stopping the treatment and continuing mowing.

4. If *Arrhenatheratum* and *Urtica* are both present, they behave oppositely under the major treatments.

5. *Urtica dioica* seems to be a rather constant species in a nitrogen rich environment with infrequent to no mowing, whereas *Lamium maculatum* seems to have an optimum during a few years after an environment has been fertilised and/or left alone for some time.

6. *Heracleum sphondylium* is an example of a species which behaves more or less independently from the major treatments while fluctuating in relation to other factors.

*Species richness*

Table 3 also shows the course of the total number of species in the six representative plots. The highest number of species found in one subplot of $5 \times 3.5$ m$^2$ is 30 (subplot 8a in 1971), the lowest number is 10 (subplot 16a in 1974). These numbers are extremes for all subplots. In most plots species richness fluctuates considerably. Still some conclusions can be drawn:

1. The species number in mown plots tends to be around 20, that of non-mown plots around 15.

2. The relatively high numbers of species in plot 8 are caused by the occurrence of weeds in the first years after sod cotting.

3. The rapid increase in *Urtica dioica* after fertilization, as in plot 16, is accompanied by a rapid decrease in species number, but the reverse relation occurs after stopping fertilization.

4. There is an indication of a correlation between the occurrence of high quantities of *Heracleum sphondylium* and relatively low species numbers in mown plots.

**Discussion**

The main aim of this contribution is to indicate major changes in the floristic composition of vegetation under different treatments and to show the perspectives of numerical methods in detecting such changes. The sug-

99

Table 3. Occurrence of four leading species and species richness in six representative experimental plots 1970–1978.

|  | PLOT 1 | | | | | | | | | PLOT 8 | | | | | | | | |
|---|---|---|---|---|---|---|---|---|---|---|---|---|---|---|---|---|---|---|
|  | 70 | 71 | 72 | 73 | 74 | 75 | 76 | 77 | 78 | 70 | 71 | 72 | 73 | 74 | 75 | 76 | 77 | 78 |
| Arrhenatherum elatius | 0 | 0 | 0 | 1 | 1 | 3 | 5 | 5 | 6 | 0 | 0 | 2 | 2 | 3 | 0 | 6 | 7 | 7 |
| Urtica dioica | 6 | 6 | 5 | 5 | 3 | 4 | 3 | 3 | 3 | 8 | 0 | 1 | 1 | 1 | 0 | 0 | 0 | 0 |
| Lamium maculatum | 6 | 7 | 6 | 3 | 2 | 4 | 3 | 3 | 3 | 4 | 0 | 0 | 0 | 0 | 0 | 0 | 0 | 0 |
| Heracleum sphondylium | 2 | 1 | 2 | 5 | 5 | 3 | 3 | 6 | 5 | 5 | 2 | 1 | 1 | 2 | 3 | 3 | 6 | 6 |
| Number of species | 24 | 18 | 20 | 18 | 19 | 26 | 19 | 23 | 23 | 24 | 30 | 25 | 24 | 23 | 28 | 20 | 20 | 21 |

|  | PLOT 2 | | | | | | | | | PLOT 13 | | | | | | | | |
|---|---|---|---|---|---|---|---|---|---|---|---|---|---|---|---|---|---|---|
|  | 70 | 71 | 72 | 73 | 74 | 75 | 76 | 77 | 78 | 70 | 71 | 72 | 73 | 74 | 75 | 76 | 77 | 78 |
| Arrhenatherum elatius | 0 | 0 | 0 | 0 | 0 | 0 | 0 | 3 | 4 | | | 0 | 0 | 0 | 2 | 0 | 0 | 0 |
| Urtica dioica | 5 | 7 | 5 | 6 | 5 | 5 | 7 | 5 | 6 | | | 3 | 5 | 5 | 4 | 5 | 6 | 7 |
| Lamium maculatum | 5 | 7 | 6 | 9 | 9 | 9 | 8 | 8 | 8 | | | 2 | 2 | 4 | 4 | 6 | 6 | 7 |
| Heracleum sphondylium | 5 | 1 | 5 | 7 | 5 | 1 | 5 | 1 | 2 | | | 5 | 7 | 6 | 4 | 2 | 3 | 5 |
| Number of species | 16 | 11 | 16 | 14 | 14 | 15 | 15 | 15 | 15 | | | 20 | 19 | 19 | 19 | 15 | 19 | 17 |

|  | PLOT 16 | | | | | | | | | PLOT 20 | | | | | | | | |
|---|---|---|---|---|---|---|---|---|---|---|---|---|---|---|---|---|---|---|
|  | 70 | 71 | 72 | 73 | 74 | 75 | 76 | 77 | 78 | 70 | 71 | 72 | 73 | 74 | 75 | 76 | 77 | 78 |
| Arrhenatherum elatius | | | 5 | 7 | 6 | 5 | 8 | 8 | 8 | | | 9 | 8 | 9 | 8 | 9 | 8 | 9 |
| Urtica dioica | | | 3 | 6 | 9 | 9 | 7 | 3 | 3 | | | 1 | 2 | 2 | 2 | 0 | 0 | 0 |
| Lamium maculatum | | | 0 | 0 | 0 | 0 | 5 | 0 | 0 | | | 0 | 0 | 0 | 0 | 0 | 0 | 0 |
| Heracleum sphondylium | | | 6 | 8 | 7 | 2 | 3 | 3 | 5 | | | 7 | 7 | 5 | 2 | 2 | 5 | 3 |
| Number of species | | | 18 | 17 | 10 | 15 | 16 | 18 | 22 | | | 13 | 16 | 13 | 24 | 12 | 18 | 20 |

gestion of the combined application of classification, ordination, species richness data and analyses of single species behaviour is that each of these approaches helps us in revealing the dynamical structure of this vegetation.

As has been shown before (e.g. van der Maarel 1969, Londo 1971, Austin 1977) ordination diagrams of site-time series in which the positions of plots in subsequent years are connected, are very elucidative. In the present case it can be directly shown how the two main groups of plots behave. The mown plots from the eastern orchard part move from a tall forb and grass vegetation with dominant *Urtica dioica* to a grassland vegetation with dominant *Arrhenatherum elatius*. It is also suggested by the ordination diagram that the plots from the eastern part, being much alike at the beginning of the observations, tend to diverge in the rate at which they converge with the *Arrhenetherum* grassland which shows a rather stable composition. On the other hand the not-mown plots tend to move more irregularly, and the direction of change cannot yet be predicted.

The classification presents a useful scheme of the species which are important in the determination of clusters and a summary of the species composition of the more or less constant plots as well as the earlier and later stages of the changing plots. From the high number of relevés in type 6a it can be assumed that a convergence towards that type is taking place. The dynamical relationships between types could be represented in the form of a kinematic graph (Londo 1971) or more quantitatively in a transition diagram (cf. Austin 1980) upon which predictions on future developments can be based. This will be done after the 1979 field data will be available.

The few data on species behaviour in time suggest the distinction between trend following, fluctuating and constant species. It would be necessary to study the entire data set more carefully and to relate changes found to possible fluctuations in weather conditions, or, as M.J.M. Oomes (priv. comm.) suggested fluctuations in the phenological date of mowing.

Data on species richness should be accompanied by

estimations of a 'heterogeneity index of diversity' (Peet 1974). The problem in our case is that the estimates of species quantities are probably too rough to permit the calculation of any index. The 1979 team of students, Kittie Klunder and Jeanne van Sebille, found that the reciprocal Simpson index (cf. Hill 1973) based on relative cover values could be applied in the detection of differences in 'heterogeneity-diversity.

The interpretation of the observed changes in terms of the availability, removal (through mowing) and cycling of nutrients is still under way. Total yields of the plots did not vary much and appeared to be in the order of 400-600 g/m² (4-6 t/ha) of dry matter. These data refer to rather dry seasons, however. As such they are very similar to the yields of fertilized plots in the Rothamsted experiment (Thurston et al. 1976). Crop and soil analyses available so far do not allow conclusions. The extensive data collected in 1978, and the planned analyses of 1979, both not yet available, will enable to study this matter in more detail. As a preliminary observation we mention the obvious reaction of Urtica dioica (as well as Lamium maculatum) on the supply of nitrogen. This reaction seems to be the same for a gift of NPK and N only. This would suggest that the Urtica dioica populations in this area primarily respond to nitrogen, where phosphorus is apparently not a limiting factor at the beginning. The obvious response of Urtica to phosphorus described for woodland populations by Pigott (1971, Pigott & Taylor 1964) seems to deviate from our observations. May be the difference is due to the phase of the species concerned: in our case the response is at least partly in the vegetative growth of mature populations, in Pigott's case the emphasis is on seedling development. Another difference is found in the light climate the populations types are adapted to. Here we would need further chemical analysis and demographic studies, which are planned for the near future.

## Summary

Vegetation development in a former orchard on loamy moist soil after removal of the fruit trees under different treatment is described with the help of multivariate methods. Data on 20 plots over 9 years are treated with a simple classification program and a PCA ordination. Two main groups of plots are involved: one in a relatively moist part of the area where the trees were removed in 1970, one in a slightly drier area where the trees had been removed earlier. The first group is characterized by forbs such as Urtica dioica and Lamium maculatum, the second by grasses, mainly Arrhenatherum elatius.

Under a mowing regime the plots of the first group change in the direction of the second group. Under continued fertilization the Urtica vegetation can be maintained. Establishment and/or rapid spread of Urtica can be brought about in Arrhenatherum dominated sites by (re-) introduction of nitrogen fertiliser.

Not mowing leads to a dense Urtica-Lamium stand in which Lamium maculatum finds an optimum some years after mowing is stopped.

The methods of summarizing the site-time data are discussed and some new ones announced for further research.

## References

Austin, M.P. 1977. Use of ordination and other multivariate descriptive methods to study succession. Vegetatio 35: 165–175.

Austin, M.P. 1980. An exploratory analysis of grassland dynamics: an example of a lawn succession. In E. van der Maarel (ed.) Advances in vegetation science: Succession. Vegetatio 43: 87–94.

Hill, M.O. 1973. Diversity and evenness: a unifying notation and its consequences. Ecology 54: 427–432.

Londo, G. 1971. Pattern and process in dune slack vegetation along an excavated lake in the Kennemer dunes (the Netherlands). Thesis Nijmegen, 270 pp.

Maarel, E. van der, 1969. On the use of ordination models in phytosociology. Vegetatio 19: 21–46.

Maarel, E. van der, 1978. Experimental succession research in a coastal dune grassland. A preliminary report. Vegetatio 38: 21–28.

Maarel, E. van der, 1979. Transformation of cover-abundance values in phytosociology and its effects on community similarity. Vegetatio 39: 97–114.

Maarel, E. van der, J. G.M. Janssen & J.M.W. Louppen, 1978. TABORD, a program for structuring phytosociological tables. Vegetatio 38: 143–156.

Peet, R.K., 1974. The measurement of species diversity. Ann. Rev. Ecol. Syst. 5: 285–307.

Pigott, C.D., 1971. Analysis of the response of Urtica dioica to phosphate. New Phytol. 70 953–966.

Pigott, C.D. & K. Taylor, 1964. The distribution of some woodland herbs in relation to the supply of nitrogen and phosphorus in the soil. J. Ecol. 52 (suppl.): 175–185.

Thurston, J.M., E.D. Williams & A.E. Johnston, 1976. Modern developments in an experiment on permanent grassland started in 1856: effects of fertilisers and lime on botanical composition and crop and soil analyses. Ann. Agron. 27: 1043–1082.

Westhoff, V. & A.J. den Held, 1969. Plantengemeenschappen in Nederland. Thieme, Zutphen, 324 pp.

Westhoff, V. & E. van der Maarel, 1978. The Braun-Blanquet
  approach. 2nd ed. In: R.H. Whittaker (ed.). Classification
  of plant communities, p. 287–399. Junk, The Hague, 408 pp.
Williams, E.D., 1978. Botanical composition of the Park Grass
  plots at Rothamsted 1856–1976. Rothamsted. Exp. Station,
  Harpenden, 59 pp.

Accepted 20 December 1979

SUCCESSION IN A SOUTH SWEDISH DECIDUOUS WOOD: A NUMERICAL APPROACH

Stefan PERSSON**

Institute of Plant Ecology, University of Lund, ö Vallgatan 14, S-223 61 Lund, Sweden

Keywords:
Classification, Ordination, Semipermanent plots, Succession

## Introduction

Studies of successions in vegetation must, in many cases, involve observation periods of considerable length. Especially when successions are dependent on or strongly influenced by developments in tree- and shrub layers, observations must be extended over decades. Very rarely one and the same investigator is in a position to be able to observe his permanent plots for such a long time. An alternative to this dilemma is the possibility of reestablishing, with acceptable accuracy, the plots once used and having them analyzed by other investigators. The approach thus uses semipermanent plots, and I use the term corresponding quadrats for those unit areas used to sample one such plot at different points in time.

This method was used by Lindgren (1971), who in 1969 reestablished 74 plots, originally laid out by Lindquist (1938) in 1935 in the Swedish national park Dalby Söderskog. In the summer of 1975 and in the spring and summer of 1976 these investigations were repeated by the author, once again using Lindquist's 74 – plot system from 1935.

The area, a mixed deciduous wood of elm, ash, oak and beech, was established in 1918 as a national park. Since then it was virtually left to itself, resulting in spontaneous successions in tree-, shrub- and field layers. The general features of the investigation area and its history were thoroughly described by Lindquist (1938) and Lindgren (1971). The development of the vegetation was analysed by Malmer, Lindgren & Persson (1978), using traditional phytosociological methods.

During the last decades numerical, computer assisted methods have been developed and have proved to be powerful tools in vegetation ecology. Such methods were also applied in studies of successions (Williams et al. 1969, van der Maarel 1969, Austin 1977). One aim of this studie is to test the usefulness and the relevance of such methods, when applied to the data available from Dalby Söderskog (cf Table 1).

## Materials and methods

### Reestablishment of the plots

The positions of the plots are given by the map and the accompanying information presented by Lindquist (1938). The baseline for the plot system is the so called straight path (Fig. 1). The plots are situated along lines, here called taxlines. at equal intervals (50 m) perpendicular to the baseline. In 1969 no original markings were found and Lindgren therefore reestablished the plots using a compass and pacing together with checks of the reasonableness of his established plots against Lindquist's vegetation data for each plot (Lindgren 1971). Lindgren marked each plot with a wooden stick. The national park is a popular recreation site and has about 25–30 000 visitors per year. Unfortunately many of the sticks were removed. In 1975 the author recovered 17 out of 74 sticks set out by Lindgren.

In the summer of 1975 the author tried to reestablish

* Nomenclature of species follows Lid (1974), except for *Crataegus oxyacantha*, and for mosses Nyholm (1954–1969). *Viola silvestris* ( = *V. reichenbachiana*) and *V. riviniana* are treated collectively as *Viola* sp. *Dactylis glomerata* includes *D. aschersoniana*, *Crataegus oxyacantha* includes *C. lagenaria*, *Eurhyngium striatum* includes *E. Zetterstedtii* and *E. praelongum* includes *E. Swartsii*.

** I wish to thank Prof Nils Malmer, Institute of Plant Ecology, Lund, for his support and valuable advice, Dr Eddy van der Maarel, University of Nijmegen, The Netherlands, for valuable advice and for providing computer programs. I also wish to thank Fil lic Lennart Lindgren and Fil kand' Tommy Wikberg for providing primary data, and Mrs Mimmi Varga for drawing the figures.

Table 1. Different data sets available from Dalby Söderskog.

| Variable(s) | Aspect | Time of sampling | Quadrat size | Reference/Collector |
|---|---|---|---|---|
| Cover, field layer species | Spring | 24/4 - 5/5 1935 | 1 m$^2$ | Lindquist (1938) |
| " | Summer | 15/8 - 25/8 1935 | " | " |
| " | Summer | 5/8 - 18/8 1969 | " | Lindgren (1971) |
| " | Spring | 10/5 - 12/5 1970 | " | " |
| " | Summer | 21/7 - 16/8 1975 | " | author |
| " | Spring | 9/5 - 14/5 1976 | " | T. Wikberg, author (unpubl) |
| " | Summer | 3/8 - 9/8 1976 | " | author |
| Cover, tree- and bush layer species | - | 15/8 - 25/8 1935 | 16 m$^2$ | Lindquist (1938) |
| " | - | 5/8 - 18/8 1969 | " | Lindgren (1971, pers comm) |
| " | - | 21/7 - 16/8 1975 | " | author |
| Cover, bottom layer species | - | 15/8 - 25/8 1935 | 1 m$^2$ | Lindquist (1938) |
| " | - | 21/7 - 16/8 1975 | " | author |
| pH in soil, 0-5 cm, prim. data | - | 1) 1970 | " | Johansson & Grahn (unpubl) |

1) Averages from four samples, viz 14/6, 13/7, 16/7 and 10/8 1970.

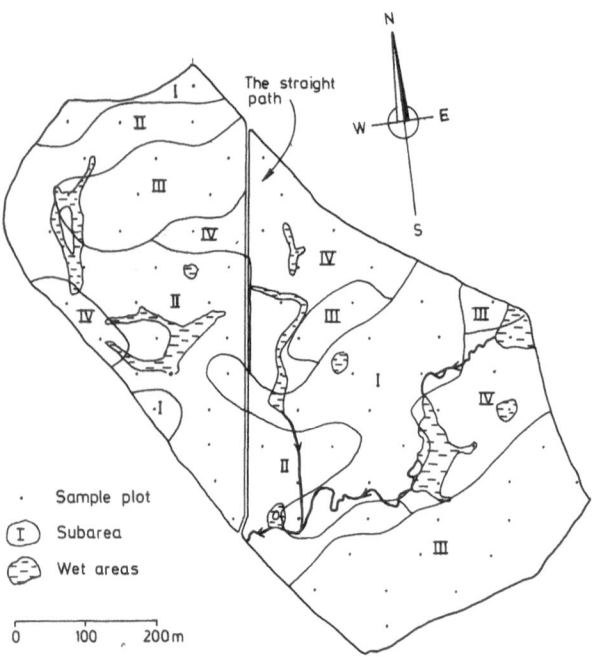

Fig. 1. Map showing Dalby Söderskog with the systematic plot system containing 74 plots (redrawn from Lindquist 1938). The plots are arranged along taxlines perpendicular to the straight path. A division into four subareas is also shown.

the plots using a compass and a decimetergraded wire. This experience showed two things: a) that Lindquist's map (cf Fig. 1) was surprisingly accurate and b) that walking by compass gave unsatisfactory angle errors when taxlines became long (300–400 m). Some of these errors could be diminished by adjustments to natural features given in the map, like footpaths and creekbends, but the usefulness of a more exact method was obvious.

In the spring of 1976 the plots were reestablished using a leveling instrument with an angular scale and an ocular which allows range finding, together with a lath. The intersections between the taxlines and the baseline were determined using a wire graded in decimeters. The levelling instrument was placed above such an intersection, the 'zero' direction defined along the baseline and the perpendicular direction established by turning the instrument 90 degrees. Thereafter the distances from an intersection to plots were found using the range finding ocular together with the lath.

Of course only the plots closest to the baseline could actually be established with the instrument positioned on the baseline. Sooner or later, when moving the lath away perpendicular from the baseline, the line of sight was cut off by a tree trunk or a shrub. It was thus necessary to move the rangefinder along the taxlines, without loosing contact with these imaginary lines and the actual distance to the baseline. This was achieved through the frequent establishment of temporary guidepoints, situated on the taxline and with a known distance to the baseline. These guidepoints were marked with easily visable yellow sticks. It was thus possible to move the range finder past these guidepoints and again place it on the taxline by adjusting its position so that, when looking 'backwards', at least two yellow sticks were in line. By measuring the distance to the nearest 'backward' guidepoint and sum up, the distance between the rangefinder's present position and the baseline was known.

When established each plot was marked with an uncoloured bamboostick in order to ensure exact correspondance in position of the quadrats in spring and summer 1976. Only 3 out of 74 of these sticks were lost by August 1976. The lost sticks were easily replaced following verbal descriptions, referring the position of each plot to pathes, conspicuous trees etc. The same correspondance between subsequent investigations in the same or the following year was achieved by the two former investigators (Lindquist 1935 p. 209, Lindgren 1971 p. 11).

*Vegetation data*

Each plot was sampled using two quadrats of different size with a mutual centre and orientated parallel to the taxlines. The field layer and the bottom layer were sampled using quadrats of 1 × 1 sq. m. The tree- and shrub layers were differentiated according to height: A1-, A2-, B1- and B2-layers, height > 15, 15–8, 8–2 and 2–0.8 m resp. (Lindquist 1938). These layers were sampled in quadrats of 4 × 4 sq. m. Individuals of tree- and shrub species, usually seedlings, lower than 0.8 m were considered as part of the field layer and thus sampled in the 1 × 1 sq. m quadrats.

In 1935 and 1969/70 cover estimates according to the Hult-Sernander - Du Rietz scale (Trass & Malmer 1978) were used as a quantitative measure of species importance, while in 1975–1976 the % cover was estimated. The percentage data were later transformed to the Hult-Sernander- Du Rietz scale, when comparisons with older data were made.

The vegetation shows distinct spring- and summer aspects. Consequently all investigations concerning the field layer have been carried out twice in the same plots, either in spring and following summer or the reverse (see Table 1). Several tests indicated that a combined treatment of the spring and summer aspects would be the most profitable approach (see also Malmer et al 1978). The field layer data from consecutive spring- and summer aspects were therefore lumped together to form a single data set. Originally a few species were present in both aspects, and therefore one of the records were deleted. In such cases the record, where the species had its maximum development, were kept and the other deleted. Of the resulting three data matrices, each consisting of 74 quadrats, the first and the third represent spring and following summer of 1935 and 1976, while the second matrix represents the summer of 1969 and the following spring of 1970. Consequently the data matrices will be referred to as 1935/35, 1969/70 and 1976/76. For some treatments these matrices were combined to one single matrix, containing 222 quadrats and 62 species (74 quadrats × 3 times × 62 species).

The field layer data from the summer of 1975 was used in Malmer et al (1978), but is not used here. Instead the data from the summer of 1976 is used, since they were obtained when using a more accurate method for reestablishing the plots, and since the correspondance between spring- and summer investigations was assured by means of markers (bamboosticks).

*Alternative quadrats*

The data in matrices 1935/35, 1969/70 and 1976/76 represent quadrats positioned in the same 74 plots at six different times. The position of a quadrat can not be expected to coincide completely between these matrices, i.e. between different investigations, since permanent markers have vanished. This causes a misplacement error when data from corresponding quadrats are analyzed. In order to evaluate this effect an alternative plot system was used in the spring and summer of 1976. This system consisted of 74 alternative quadrats, each situated 5 m north of the corresponding 'real' or reestablished quadrat. The field experiences indicated that a deviation of this magnitude was possible but not evident. The data concerning spring- and summer aspects from the alternative quadrats were also lumped together to a single matrix, here referred to as 1976/76- Alt.

*Environmental data and ecological indicator values*

Lindquist (1935) presented some environmental data from the systematic plot system, viz. pH, calcium-, phosphate-, potassium- and nitrate concentrations in soil samples. in the form of maps with different-sized circles. Johansson & Grahn (unpubl.) studied the changes in chemical variables, primarily pH, by comparison with data derived from Lindquist's maps. Their own data from the plots established by Lindgren in 1969–70 are available (cf Table 1).

In situations where environmental data are scarce or absent, changes in the environment can often be deduced from changes in the vegetation, by means of known properties of the species involved. To explore this possibility the field layer data were combined with ecological indicator values for the species according to Ellenberg (1974) to a synthetic characteristic indicator value (CIV) for each quadrat. The environmental factors estimated in this way are derived from the occurrence of species in relation to (Ellenberg op cit):

a) relative light intensity during summer time.
b) the degree of continentality of the general climate
c) soil moisture or water level
d) the ammonium or nitrate supply
e) soil reaction

The CIV's are calculated as the sum of products between cover value, in this case according to the Hult-Sernander-Du Rietz scale, and indicator value for a species, divided by the sum of 'involved' cover values, summations performed over all species in a quadrat. Those indicator values denoted as 'X' or '?', representing indifferent or unknown relationship of a species to an environmental factor, or lacking in Ellenberg (1974) are assigned zeros. The zeros represent unknown values and can therefore not contribute to the sum of products, and consequently the associated cover values are excluded from the sum of cover values, they are not 'involved'.

The CIV's are interpreted as rough estimates, on an arbitrary scale, of the intensity of an environmental variable in a quadrat, as revealed by the species present in that quadrat. The prerequisites for using characteristic indicator values as estimates of prevailing and former environmental conditions are in the present case somewhat disadvantageous. This is because the wood is very uniform, with a few very dominant species, and because the realized amplitude of many environmental factors is rather narrow. The number of species in each plot that can contribute and add precision to a calculated characteristic indicator value is furthermore rather limited. In spite of this an acceptable regression is found between the actual pH measurements in each plot from 1970 (Johansson & Grahn, unpubl) and the CIV's for reaction status, calculated from submatrix 1969/70 (Fig 2).

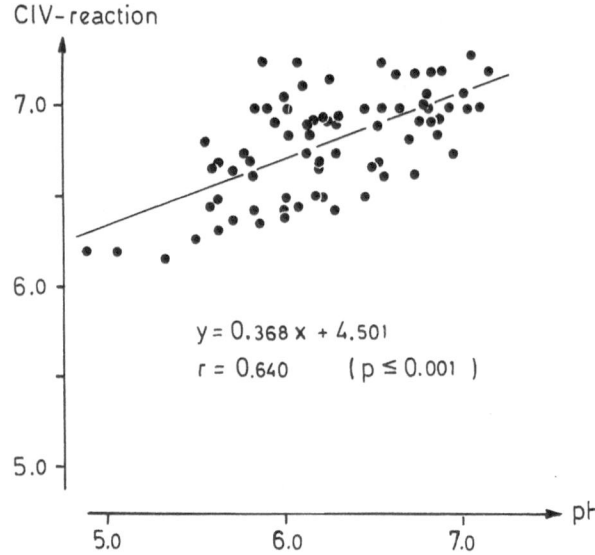

Fig. 2. The regression between characteristic indicator values (CIV's) for reaction status calculated from the 1969/70 submatrix and the actual pH values in soil samples from each plot (1970). The latter data are due to Johansson & Grahn (unpubl.).

*Numerical methods*

Classifications were obtained using the TABORD-program (Janssen 1972, Persson 1977, van der Maarel et al 1978), which performs an agglomerative clustering of quadrats and produces a structured phytosociological table of these clusters or groups. The program requires an initial classification to work with. This was achieved through the divisive, monothetic information drop method of Lance & Williams (1968, program DIVINF).

Ordinations were obtained using the program ORDINA (Roskam 1971), which performs a centered, non-standardized principal component analysis by extracting principal components from a matrix of Euclidean distances between quadrats.

Diversity indices are normally calculated from the number of individuals of species, a type of data not available in the present study. McIntosh (1967, formula 4) proposes however an index based on a model of distances between quadrats and the origin in a coordinate system in a multidimensional species space, with as many axes as there are species. McIntosh's index is in this study applied to a species space, where the quantitative scores on the axes are the cover values for each species in a quadrat, instead of the number of individuals. Thus the sample size (total number of individuals) instead becomes the sum of cover values. The diversity index was calculated for each quadrat and proved effective in dealing with the two aspects of diversity, viz the number of species in a quadrat and the evenness, i.e. to what extent those species share the space equally or have equal cover values.

## Results

### Classification

The material was classified on the basis of field layer species by applying the TABORD-program to the data from all three investigations, i.e. a matrix of 222 quadrats

Table 2. Classification of the 222 × 62 matrix using the TABORD-program. Species names are abbreviated according to the RUBIN-system (Österdahl et al 1977), and the full species names are listed in the Appendix. Sections are delimited by thick solid lines. Each group is subdivided according to the year of sampling (broken lines).

Table 2. (cont.).

|  | SECTION III | | | | | | SECTION IV | |
|---|---|---|---|---|---|---|---|---|
| TABLE PART 2 | | | | | | | | |
| YEAR | -76 | -35 | -69 | -76 | -35 | -69 | -76 | -35 | -69 | -76 |
| GROUP | | | | | | | | |

| | | | | | | | |
|---|---|---|---|---|---|---|---|
| ADOX MOS | | | | | | | |
| CHRY ALT | | | | | | | |
| GAGE MIN | | | | | | | |
| CORY FAB | | | | | | | |
| LAMI GAL | | | | | | | |
| FRAX EXC | | | | | | | |
| GAGE SPA | | | | | | | |
| GAGE LUT | | | | | | | |
| AEGO POD | | | | | | | |
| ANEM NEM | | | | | | | |
| ANEM RAN | | | | | | | |
| MERC PER | | | | | | | |
| RANU FIC | | | | | | | |
| CORY BUL | | | | | | | |
| GEUM RIV | | | | | | | |
| FILI ULM | | | | | | | |
| CIRC LUT | | | | | | | |
| URTI DIO | | | | | | | |
| GEUM URB | | | | | | | |
| ACTA SPI | | | | | | | |
| ANGE SIL | | | | | | | |
| ANTH SIL | | | | | | | |
| C REMOTA | | | | | | | |
| C SILVAT | | | | | | | |
| CALT PAL | | | | | | | |
| CAMP LAT | | | | | | | |
| CAMP TRA | | | | | | | |
| CORY AVE | | | | | | | |
| CRAT OXY | | | | | | | |
| CREP PAL | | | | | | | |
| DACT GLO | | | | | | | |
| EPIL MON | | | | | | | |
| EPIP HEL | | | | | | | |
| EQUI SIL | | | | | | | |
| FAGU SYL | | | | | | | |
| FEST GIG | | | | | | | |
| FRAG VES | | | | | | | |
| GALI APA | | | | | | | |
| GERA ROB | | | | | | | |
| LATH VER | | | | | | | |
| LONI XYL | | | | | | | |
| OXAL ACE | | | | | | | |
| PARI QUA | | | | | | | |
| POA NEM | | | | | | | |
| POA TRI | | | | | | | |
| POLY MUL | | | | | | | |
| PULM OFF | | | | | | | |
| QUER ROB | | | | | | | |
| RANU AUR | | | | | | | |
| RANU REP | | | | | | | |
| RUME SAN | | | | | | | |
| SANI EUR | | | | | | | |
| ST NE.GL | | | | | | | |
| STAC SIL | | | | | | | |
| TARAXACZ | | | | | | | |
| ULMU GLA | | | | | | | |
| VERO CHA | | | | | | | |
| VERO HED | | | | | | | |
| VERO MON | | | | | | | |
| VIBU OPU | | | | | | | |
| VICI SEP | | | | | | | |
| VIOLAZ | | | | | | | |

Anemone nemorosa- A. ranunculoides community
Anemone ranunculoides - Mercurialis perennis dominated variant

Ranu fic community
Cory bul variant

Ranu fic com. Geum riv-Fili ulm var.

Ranu fic com. Urtica dioica-variant

and 62 species. This approach is equivalent to the site-period analysis recommended by Williams et al (1969), though they used another classification method. The similarity ratio was used as a similarity coefficient (Wishart 1969, Westhoff & van der Maarel 1978). The required initial classification contained 12 initial groups and was obtained by applying the DIVINF-program to the 222 × 62 material.

In the final classification (Table 2) the following optional parameters were used: No groups with less than three members were allowed, a threshold value of 0.3 was used to prevent incorporation of quadrats with a similarity less than 0.3 to a group and a fusion limit of 0.85 preventing fusion of groups with less mutual similarity. Constant species were defined as species with at least 50 % frequency in a group.

The classification is based on synthetic quadrats, each made up of two different populations replacing each other in time, viz. the spring- and summer aspects, but located in the same quadrat. The high fusion limit was chosen to prevent an unwanted fusion of the two largest groups, 2 and 7, at the similarity level of 0.832. The main difference between those groups consists of a difference in balance in the cover of the two codominating vernal species *Anemone nemorosa* and *A. ranunculoides*. In group 2 the former is generally dominating, while in group 7 the latter obtains the higher cover estimates. Though *Mercurialis perennis* is strongly dominating in the summer aspects of both groups, it has a higher cover in group 7. This difference is felt to be of significance in the present study, and thus the fusion of the two groups was prevented. This leads to a classification and division on a rather detailed level. The aim is to study vegetation changes, and the level of division thus has to correspond with the level, where such changes take place.

The sequence of groups in the phytosociological table, constructed by the program, was not optimal. The sequence was therefore altered, according to an analysis of similarities between groups (Table 3), to the sequence presented in Table 2. This sequence reflects the major vegetational gradient in the data set, along which *Anemone nemorosa*, *A. ranunculoides* and *Ranunculus ficaria* gradually replace each other as dominants in the spring aspect.

Among the 11 terminal groups 4 main groupings or sections (I–IV) can be distinguished, in order to demonstrate the successional relations in Table 2. In section I all groups but group 10 are either entirely made up, or at least dominated, by quadrats from 1935 (Table 2).

Of the 63 quadrats in section 1, 67 % originate from 1935, 16 % from 1969/70 and 17 % from 1976. In section III both groups are dominated by quadrats from 1976. Of the 76 quadrats in this section 9 % originate from 1935, 34 % from 1969/70 and 57 % from 1976. In the intermediate section II 29 % of the 63 quadrats originate from 1935, 49 % from 1969/70 and 22 % from 1976. The last section (IV) is a constant feature containing the quadrats situated in the wet depressions, where ca. one third of its 20 quadrats originate from each period (35, 35 and 30 % respectively). The group sequence in Table 2 is thus reflecting a successional gradient, represented by sections I–III, where the origin of the majority of quadrats in each section is shifting from 1935 over 1969/70 to 1976 (cf Table 6).

The quadrat groups in Table 2 can be interpreted as representing two levels in a hierarchical system, viz communities and variants of communities (Table 3). The terms are used here as neutral concepts, without any connexion to the levels of any established phytosociological hierarchy. The vegetation units are named according to dominants of the spring- and summer aspects.

The classification is basically structured into four communities, differentiated by some vernal species (Table 2). In two of the comm. *Anemone nemorosa* is clearly dominating. The *Anemone nemorosa – Adoxa moschatellina* – community (group 4) was only found in 1935 and contained some species with a restricted distribution in both space and time, e.g. *Adoxa moschatellina*, *Chrysosplenium alternifolium*, *Geum urbanum*, *Gagea minima*

Table 3. Nomenclature and hierarchy of groups from the classification. Similarities between groups. The figures given are the similarity ratio * 1000, selfcomparisons omitted. The three largest values in each row are underlined, provided they exceed 0.5.

| | Group number | 4 | 5 | 10 | 1 | 9 | 2 | 7 | 3 | 8 | 6 | 12 |
|---|---|---|---|---|---|---|---|---|---|---|---|---|
| | Size: | 15 | 14 | 12 | 13 | 9 | 63 | 49 | 27 | 3 | 13 | 4 |
| Anemone nemorosa – Adoxa moschatellina community | 4 | – | 315 | 279 | 340 | 402 | <u>518</u> | 368 | 277 | 238 | 87 | 83 |
| Anemone nemorosa – Lamium galeobdolon community | 5 | 315 | – | <u>529</u> | <u>548</u> | 489 | <u>540</u> | 331 | 266 | 200 | 131 | 163 |
| Anemone nemorosa – A. ranunculoides community | | | | | | | | | | | | |
|    Fraxinus (juv) variant | 10 | 279 | <u>529</u> | – | <u>561</u> | 506 | <u>652</u> | 507 | 284 | 180 | 56 | 44 |
|    Gagea spathacea variant | 1 | 340 | 548 | 561 | – | <u>585</u> | <u>652</u> | 486 | <u>607</u> | 532 | 343 | 335 |
|    Aegopodium podagraria variant | 9 | 402 | 489 | 506 | <u>585</u> | – | <u>682</u> | <u>657</u> | 548 | 503 | 163 | 200 |
|    A. nemorosa dominated variant | 2 | 518 | 540 | <u>652</u> | <u>652</u> | <u>682</u> | – | <u>832</u> | 586 | 418 | 107 | 126 |
|    A. ranunculoides dominated variant | 7 | 368 | 331 | 507 | 486 | <u>657</u> | <u>832</u> | – | <u>666</u> | 503 | 93 | 131 |
| Ranunculus ficaria community | | | | | | | | | | | | |
|    Corydalis bulbosa variant | 3 | 277 | 266 | 284 | 607 | 548 | <u>586</u> | <u>666</u> | – | <u>724</u> | 312 | 400 |
|    Intermediate variant | 8 | 238 | 200 | 180 | <u>532</u> | 503 | 418 | 503 | <u>724</u> | – | <u>506</u> | 420 |
|    Geum rivale – Filipendula ulmaria var | 6 | 87 | 131 | 56 | 343 | 163 | 107 | 93 | 312 | <u>506</u> | – | <u>561</u> |
|    Urtica dioica variant | 12 | 83 | 163 | 44 | 335 | 200 | 126 | 131 | 400 | 420 | <u>561</u> | – |

109

and *Corydalis fabacea*. The two latter species were by 1976 not found at all in any quadrat, while the others have decreased drastically. *Gagea minima* was however recorded outside the plot system on the very edge of the wood, facing open grazed areas. The occurrences of *Filipendula ulmaria* and *Chrysoplenium alternifolium* in this group indicate a relationship with the groups 6 and 12.

In the *Anemone nemorosa* – *Lamium galeobdolon* – comm. (group 5) some low frequent species have a characteristic occurrence, e.g. *Oxalis acetosella*, *Stellaria nemorum* ssp *glochidisperma*, *Veronica montana*, *Vicia sepium* and *Fagus* saplings. *Lamium galeobdolon* is dominating in the summer aspect.

The two *Anemone* species are codominating in the *Anemona nemorosa* – *A. ranunculoides* – comm. Five variants can be distinguished:

Group 10: characterized by *Fraxinus excelsior* saplings (< 0.8 m in height).

Group 1: characterized by *Gagea spathacea* and *G. lutea*. The high cover values for *Ranunculus ficaria* indicate a relationship with the *Ranunculus ficaria*-community.

Group 9: characterized by *Aegopodium podagraria* and *Gagea spathacea*.

Group 2: where *Anemone nemorosa* generally has a higher cover than *A. ranunculoides*. The decrease in the number of species occurrences in the course of time is clearly demonstrated in this group. The average number of species per quadrat in the group was 7.5 in 1935, 5.1 in 1969/70 and 4.8 in 1976.

Group 7: where *A. ranunculoides* always dominates over *A. nemorosa* in the spring and where *Mercurialis perennis* has its maximum development during the summer.

The fourth, *Ranunculus ficaria* comm., is in spring dominated by this species and further differentiated in four variants:

Group 3: with considerable quantities of *Corydalis bulbosa* and *Aegopodium podagraria*.

Group 6: characterized by *Geum rivale* and *Filipendula ulmaria*, and where *Mercurialis perennis* is almost absent. Other more or less characteristic species are *Chrysosplenium alternifolium*, *Circaea lutetiana*, *Urtica dioica*, *Geranium robertianum*, *Poa trivialis*, *Ranunculus repens* and *Rumex sanguineus*.

Group 8: transitional between the two former groups, e.g. *Mercurialis* is present. This is the smallest group, represented by only one quadrat from each sample.

Group 12: characterized mainly by the high cover of *Urtica dioica* and the complete absence of *Anemone* species.

A separate classification based on data from tree- and shrub layers in the 4 × 4 sq. m quadrats was attempted, but failed due to the inappropriate size of these quadrats for such purposes.

*Ordination*

From Table 2 it is evident, that the variation in the material is more than one-dimensional, but in a one-dimensional sequence of groups, only one vegetational gradient can be successfully represented. This multidimensional variation, as well as the reasonableness of the groups themselves, have been studied in ordinations. The available version of the ORDINA- program has a limited capacity and can only handle the material from a single year (spring and summer). Thus three separate ordinations had to be performed (Fig 3). For convenient comparisons the 1st dimension in Figs 3a and 3b are inverted, that is all coordinates are multiplied by −1. In the scatter diagrams each quadrat is labelled with the corresponding group number from the classification.

In the ordination of the 1935/35-data (Figs 3a and 3b), the program extracts three dimensions before a reasonable separation of the groups from the classification is achieved. Those three dimensions account for an extracted variance of 55.3 %. In the ordinations of the 1969/70- and 1976/76-data (Figs 3c and 3d), essentially the same configuration is realized with the 1st and 2nd dimensions, as was with the 1st and 3rd dimensions in the 1935/35-data. The extracted variances are almost the same, 54.6 and 56.0 % resp., but extracted with only two dimensions instead of three. It is obvious, that the variation in the vegetation is reduced with time in such a way, that the number of dimensions needed for a good representation in multidimensional space is reduced by one. This reduction is mainly attributable to the absence of quadrats assignable to groups 1 and 4 in the 1969/70- and 1976/76 materials.

The ordinations strengthen the classification presented. The established quadrat groups appear reasonable and rather homogenous in the scatter diagrams. The sequence of groups in Table 2, which reflects both a vegetational

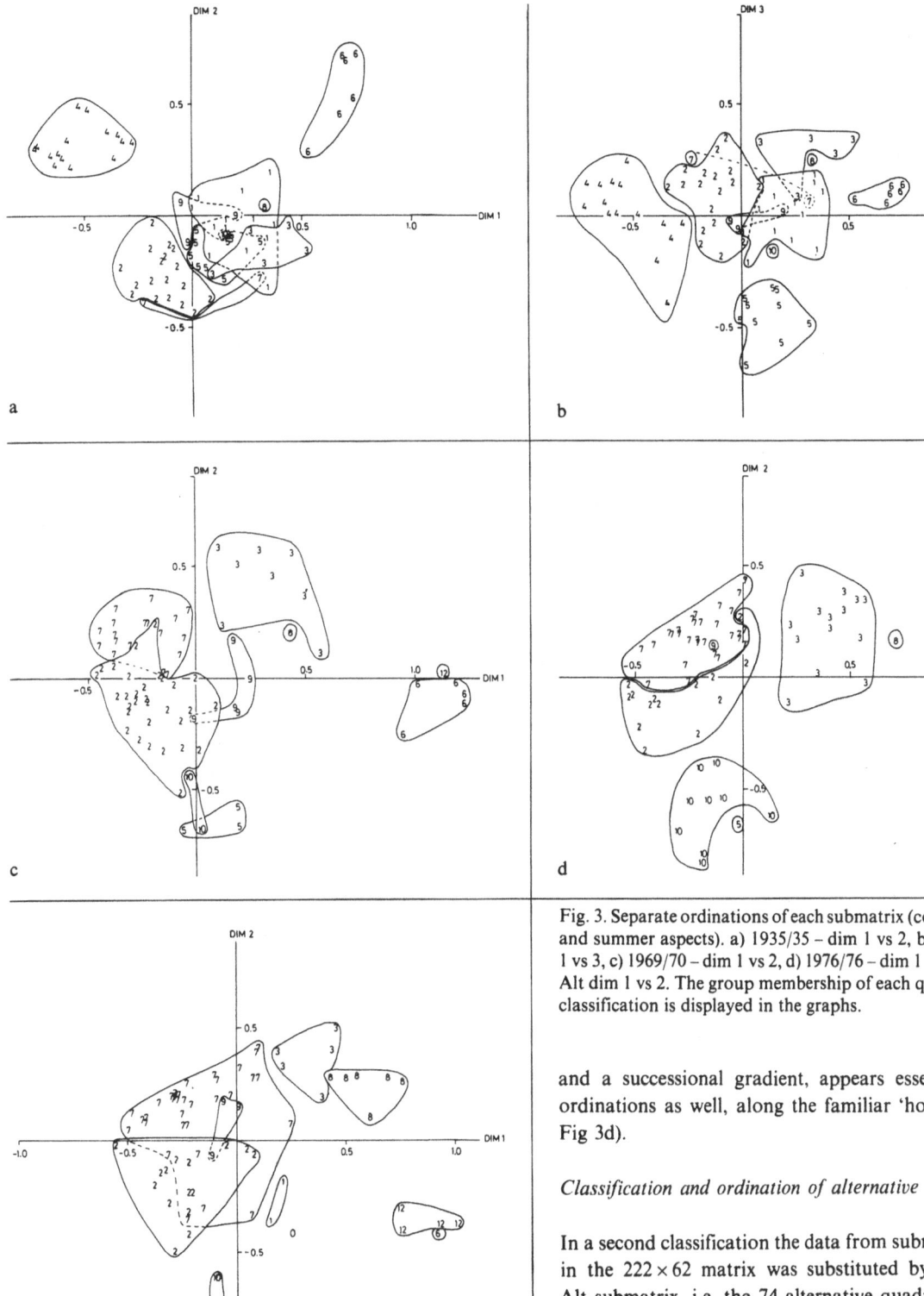

Fig. 3. Separate ordinations of each submatrix (combined spring- and summer aspects). a) 1935/35 – dim 1 vs 2, b) 1935/35 – dim 1 vs 3, c) 1969/70 – dim 1 vs 2, d) 1976/76 – dim 1 vs 2, e) 1976/76-Alt dim 1 vs 2. The group membership of each quadrat from the classification is displayed in the graphs.

and a successional gradient, appears essentially in the ordinations as well, along the familiar 'horse shoe' (e.g. Fig 3d).

*Classification and ordination of alternative quadrats*

In a second classification the data from submatrix 1976/76 in the 222 × 62 matrix was substituted by the 1976/76-Alt submatrix, i.e. the 74 alternative quadrats were used to represent the spring and the summer of 1976. This

111

new 222 × 62 matrix was then classified in order to answer the question: Would the results and conclusions be consistent if the data from the alternative quadrats were used? This means repeating the original classification on a matrix were 2/3 of the data are unchanged. The final group structure from the original classification was used as initial classification (input) and all input parameters were kept identical.

This 'experiment' had a very slight effect on the group assignment of the 148 quadrats from 1935/35 and 1969/70, only seven of these appeared in different groups compared to the original classification. Furthermore the change was in five cases only one 'step' along the linear arrangement in Table 2. All groups in Table 2 reappeared in the new classification, possessing the same characteristics and with only slightly different group sizes compared to the original classification. Of the 74 alternative quadrats 40 were assigned to equivalent groups compared to their counterparts in the original classification. One quadrat was placed in the residue (group 0), since it did not reach a similarity of 0.30 with any group.

The differences in results, when a displacement is introduced, are illustrated in Fig 3e, where an ordination of the 1976/76-Alt matrix is combined with the above mentioned classification. When this graph is compared to Fig 3d, showing the original results, it is clear that the same overall pattern of variation and community structure in the field layer emerges from the slightly misplaced sample. The number of quadrats assigned to the two adjacent groups 3 and 8 differs, but taken together that part of the diagram contains roughly the same number of quadrats in the two samples. Group 1 is appearing again in the alternative sample (cf Fig 3a-b), group 7 is somewhat extended and group 10 is less represented. These differences do not affect the general pattern of change since 1935. It can therefore be concluded, that the comparability of the samples from the original and the two reestablished plot systems are not seriously violated by misplacement errors, a feature intrinsic to the use of semipermanent plots.

*Development in groups of corresponding quadrats*

The treatments presented so far did not involve a connexion of the single quadrats, which represent the same plot at different times. The overall amount of change between the different samples or submatrices can be summarized by calculating the similarity between each pair of corresponding quadrats, and plotting the frequencies in a number

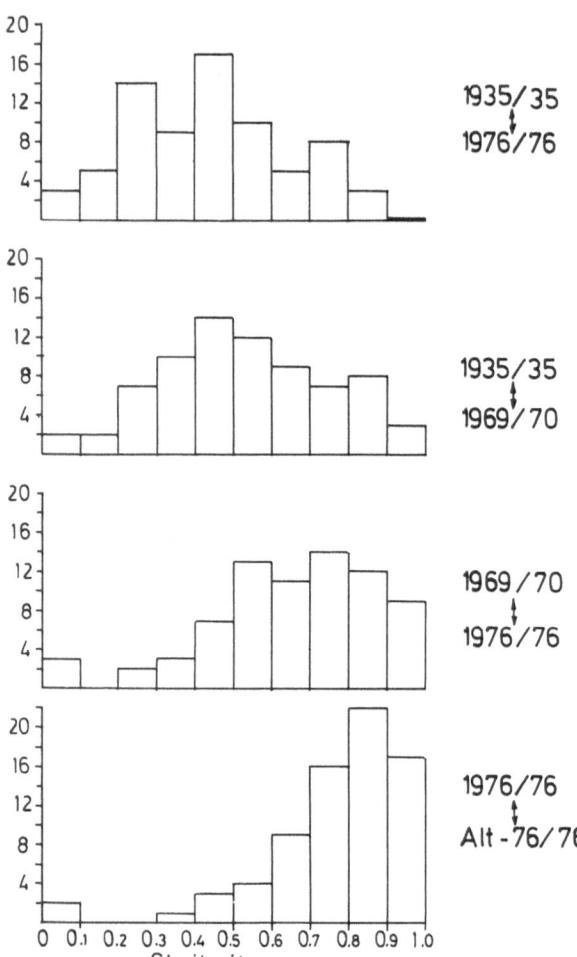

Fig. 4. Distribution of similarity ratios between pairs of corresponding quadrats, illustrating the relative degree of change between different submatrices.

of similarity classes (Fig 4). If the vegetation in Dalby Söderskog is assumed to undergo a directed change or succession since 1935, if the change started rather early and if the change is roughly proportional to the time elapsed, it can be expected that the greatest changes are found when comparing matrices 1935/35 and 1976/76, representing a time gap of 41 years. The frequency distribution of similarities between corresponding quadrats should have a peak for relatively low similarities. A slightly smaller change, and consequently a slight shift in distribution towards higher similarities, should be found when comparisons are made over 35 years, between matricies 1935/35 and 1969/70. An even smaller change, and a greater shift in distribution towards higher values, should occur when comparisons are made over only 6 years, from

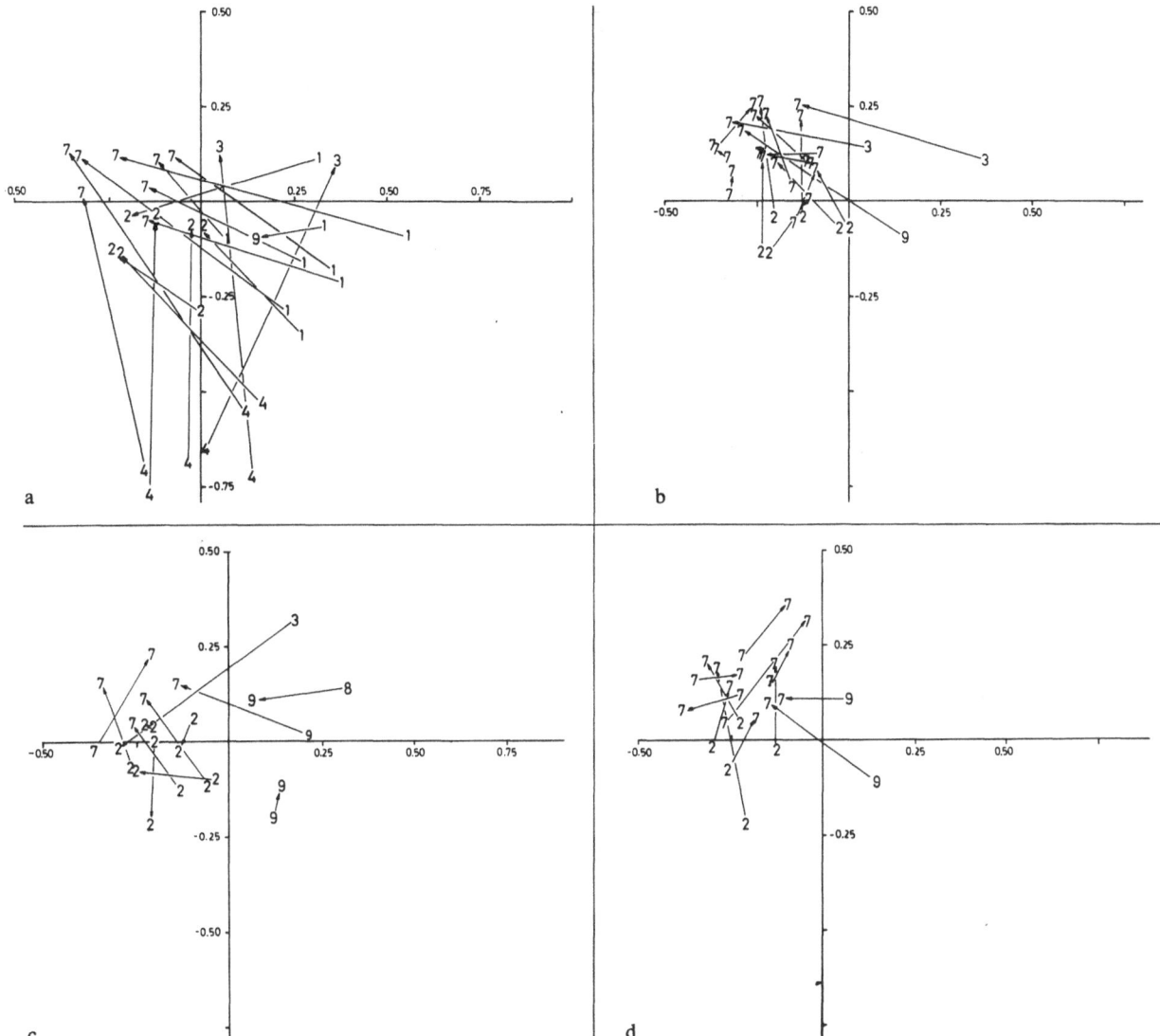

Fig. 5. Ordination of corresponding quadrats. Selection based on quadrats from 1976/76 in group 7 (Table 2). a) and c): changes between 1969/70 and 1976/76. The split on two subsets is made for convenient presentation.

1969/70 to 1976/76. Finally, when the two alternative sets, 1976/76 and 1976/76-Alt, are compared, the frequency distribution of pairwise similarities should be strongly skewed towards values near 1.0. The actual results show, although most classes are present in all comparisons, that the predicted shift in frequency distribution from low to high similarity values actually occurs (Fig. 4).

Once some quadrats were similar enough to be members of the same group, and the question is: In what way did the vegetation in such an assemblage of quadrats change

and can a pattern be recognized? This question is evaluated through an analysis of the development of the vegetation in smaller groups of corresponding quadrats. The most interesting parts of Table 2 are the groups containing quadrats from 1935 only, viz groups 4 and 1, and the expanding groups, viz groups 7 and 3. There is a strong relationship between these pairs of groups, since 21 out of the 28 quadrats corresponding to groups 4 and 1 are by 1976 classified as members of either group 7 or 3.

If the same set of quadrats are selected from the three different periods, the development through time in

113

corresponding quadrats can be studied in an ordination. This method of ordinating quadrats as time series was found useful by van der Maarel (1969) and Austin (1977). The aim of such a treatment is to discern whether a majority of quadrats show parallel or similar time trajectories through the species space of the ordination. If this is the case a general successional trend is evident (Austin op cit).

This approach was applied to one set of corresponding quadrats, viz. the 1976/76 quadrats of groups 7 and 3, and their corresponding quadrats from 1935/35 and 1969/70. A total of 129 quadrats (29 + 14 times 3) were thus ordinated in a single treatment (Fig. 5 and Fig. 6). Fig. 5 indicates a converging development for many quadrats, especially those of groups 1 and 4, towards the vegetation type of group 7, in several cases via the intermediate type of group 2. Even within group 7 itself the directed change continues in later years (Figs 5b and 5d). A similar trend is indicated in Fig. 6, where especially quadrats belonging to groups 4 and 2 develop towards the vegetation type of group 3.

These successional trends ask for attention: what specific species are responsible for the changes observed in the ordination? This question is evaluated by a tabular arrangement of sets of corresponding quadrats. Apart from the field layer, data for tree and shrub layers as well as bottom layer are included (Table 4 and 5). Only the more abundant species are presented in these tables. The characteristic indicator values, rounded off to integers, for each quadrat are also included. The CIV's are calculated over all

species though. The first selection is based on groups 4 and 1 (Table 4). Since groups 5 and 10 show a relationship, these two were treated so that all quadrats that at any time were assigned to one of these groups, were selected for comparison (Table 5). In the following discussion increases and decreases refer to changes in both frequency and cover as judged from the individual estimations of cover according to the Hult-Sernander-Du Rietz scale.

The data from corresponding quadrats, selected on the basis of group 4 show a drastic decrease in *Adoxa moschatellina*, *Chrysosplenium alternifolium*, *Gagea minima* and *Corydalis fabacea* from 1935 to 1976 (Table 4). Several other species have decreased, though less drastic, including *Anemone nemorosa*, *Geum urbanum* and *Filipendula ulmaria*. These decreases are parallelled by a considerable increase especially in *Anemone ranunculoides*, but also in *Corydalis bulbosa* and *Ranunculus ficaria*. The data from tree- and shrub layers of the same plots (16 sq. m. quadrats, years 1935, 1969 and 1975) show a decrease in *Quercus robur* (A1) and *Corylus avellana* (B1) parallelled by an increase of *Ulmus glabra* in all four layers (A1, A2, B1 and B2). Also *Fagus sylvatica* and *Fraxinus excelsior* show an increased abundance in the A1-layer of these plots. For the bottom layer species a decreasing trend is indicated, though data from 1969 are lacking.

The selection of corresponding quadrats based on group 1 in Table 2 shows decreases for field layer species such as *Gagea spathacea*, *G. lutea*, *Geum urbanum*, *G. rivale* and *Ranunculus ficaria* (Table 4). This coincides with an in-

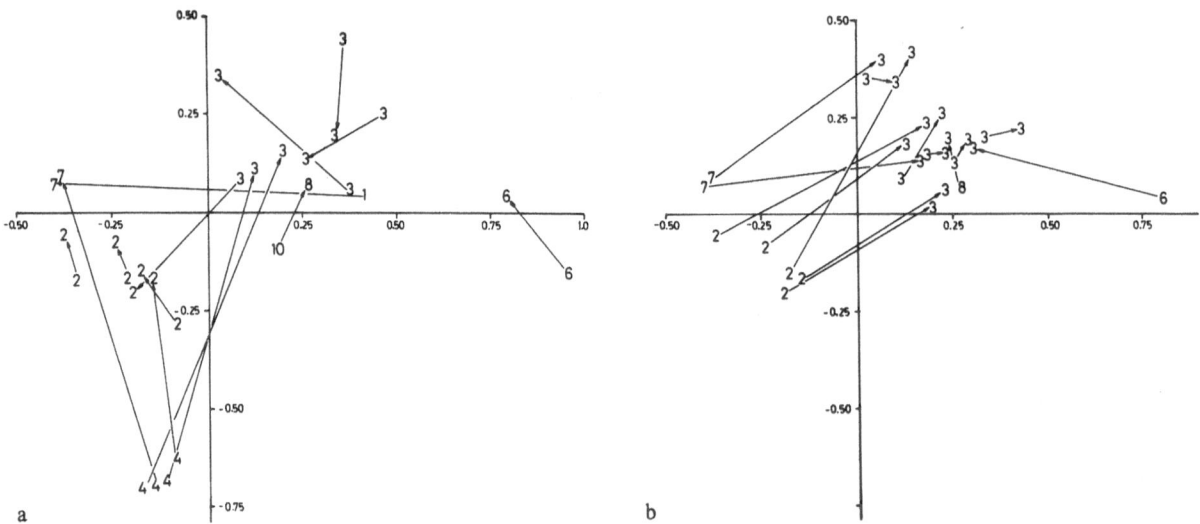

Fig. 6. Ordination of corresponding quadrats. Selection based on quadrats from 1976/76 in group 3 (Table 2). a): changes between 1935/35 and 1969/70, b): changes between 1969/70 and 1976/76.

Table 4. Development in two groups of corresponding quadrats, selected on the basis of groups 4 and 1 in Table 2.

| | 1935/35 | 1969/70 | 1976/76 | 1935/35 | 1969/70 | 1976/76 |
|---|---|---|---|---|---|---|
| | | 1 1 | 1 | 1111111111111 | | 1 |
| Group number | 444444444444444 | 772020272233233 | 737072273273737 | 1111111111111 | 7772277779727 | 2777277737770 |
| Field layer species: | | | | | | |
| ANEM RAN | --------------1- | 553331242331311 | 54445455-452414 | 3---111322135 | 4453333452444 | 2353552515454 |
| CORY BUL | --------------1 | -1---1-1-3-4345 | ----1-1153-1124 | -----------11 | -------------2 | ------------- |
| MERC PER | 35325-355435455 | 554141555545352 | 555-43444455555 | 2-21311211211 | 5553354353443 | 5555355555551 |
| RANU FIC | --1-----1-1---- | ----1---2-32232 | 143-2---1--5141 | 3321422253544 | -112-1---2121 | 22-2-2-13111- |
| AEGO POD | ----------12--- | --1-------21124 | -22-1------1-1- | --------1-1-- | ---21----121- | -----------1- |
| ADOX MOS | 345445123451551 | --------------- | --3----------- | ------------- | ----------11-- | ------------- |
| GAGE MIN | 1--121132244-45 | --------------- | --------------- | ----------11- | ------------- | ------------- |
| CHRY ALT | -3-231342213325 | --------------- | --------------- | ------------- | ------1--1--- | ------------- |
| CORY FAB | --1-1---1112111 | --------------- | --------------- | ------------- | ----------1-- | ------------- |
| ANEM NEM | 453354555545443 | 234435435521511 | 1-342543-411211 | 2144433433523 | 3234322132152 | 522242-1-2211 |
| GAGE SPA | -11--------111 | --1-1--------- | --1---------- | 121--121111-1 | ---1-11--11-- | ------------- |
| GEUM URB | --34241---1---- | ---111--------- | ----1-------1-- | -211--11--1-- | ---22--1-1--- | ------------- |
| FILI ULM | 3-1--2---2---1 | --------------- | --------------- | ------------- | ------------- | ------------- |
| GEUM RIV | 1----------1--1 | --------------- | --------------- | --1--31-2--- | ---------1--- | ------------- |
| GAGE LUT | ----11--------- | --------------- | --------------- | --1-111-131- | -----1----2-- | ------------- |
| STAC SIL | ---------1---12 | --------------- | --------------- | --1-------- | ------------- | ---1--------- |
| Tree and bush layer species: | | 1969 | 1975 | 1935 | 1969 | 1975 |
| ULMU GLA A1 | --------5555-55 | 1---21--5445-5- | -15-4---5415453 | ---5------554 | -----11---344 | -255-5--3--55 |
| ULMU GLA A2 | --------5--5--- | -532-----4----4 | 55-4----455---4 | 5-552---5---5 | -4-332--5-431 | -5-----3--343 |
| ULMU GLA B1 | 1-2--1--455-3-- | --553--533124522 | 2-451-544244143 | -4-3-5-----55 | 241--2-3--321 | 54445--4-55-5 |
| ULMU GLA B2 | ---------3--2--- | -----1------2-- | -----413141-23- | -----2------3 | -------3-3--1 | ----21-151-11 |
| FAGU SYL A1 | ------55-------- | ---3-545-------- | ---5-555--5---- | ------5------ | -------4------ | ------5------ |
| FRAX EXC A1 | --5--------5---- | 2-423-----3---- | 54--4------1-4-- | ---5-4-------5 | --4115-2-41-3 | -4--5-----55-- |
| QUER ROB A1 | 55--5--------5-- | 33--3------4-- | ----4--------- | 545----5------ | 45-----3------ | 421----4----- |
| CORY AVE B1 | --555-5----5555 | ----5-1-------4 | ----5--------2- | -2----55--5-- | -2--4--4---22 | 1--34-4--3--- |
| FRAX EXC A2 | ---542-----4--- | --------------- | --------------- | -5--5-------- | --21-----1--- | ------------- |
| FRAX EXC B1 | 5----------5- | --------------- | --------------- | 21----2--54-- | ------------- | ----------2-- |
| CORY AVE B2 | --------------- | 2-1----------- | ------------- | --------------- | -------12-1--- | ----22-1--2-- |
| CRAT OXY B2 | ----------2---- | --1-11--------- | -----1-------- | 234--232-3--- | -2--12-1-1--- | 33--14-3-1--- |
| Bottom layer species: 1935 | | | 1975 | 1935 | | 1975 |
| BRAC RUT | --------------- | ? | --------------- | -2---2---1--- | ? | ------------- |
| EURH PRA | 11-------1-42-51 | ? | 1-1-2--------- | 3-3-2522113-1 | ? | --1111-1-211- |
| EURH STR | ---------4--1--1 | ? | --2----------- | ----3--214--- | ? | ----1-------- |
| MNIU UND | -1----------1- | ? | --------------- | -----1-1-1--- | ? | ------------- |
| Characteristic Indicator Value, based on field layer species only: | | | | | | |
| CONTINENTAL (K) | 444444333443443 | 333333333333333 | 334333334333434 | 4434334434333 | 3333433434434 | 3333333?33334 |
| LIGHT FIGURE (L) | 544434434444434 | 333333233333334 | 333333333333333 | 3434434344343 | 2334323334333 | 3333332?33333 |
| MOISTURE (F) | 766666776666667 | 666666666666666 | 666666666666666 | 6666766666666 | 6666666666666 | 6666666666666 |
| NITROGEN (N) | 677777667776776 | 887777787877878 | 778878888887778 | 7776776776777 | 7777777887778 | 7787877777788 |
| REACTION (R) | 667676666676777 | 777776676777777 | 777777778777777 | 7766667777777 | 7776777777777 | 6777777777777 |

crease particularly in *Mercurialis perennis* but also in *Anemone ranunculoides*. *Ulmus glabra* is increasing in the two shrub layers (B1 and B2), while *Quercus robur* (A1) and *Corylus avellana* (B1) are rather constant. The moss layer seems to be slowly decreasing.

The selection of corresponding quadrats based on groups 5 and 10 indicates a slightly different development in time (Table 5). There is a drastic decrease in *Lamium galeobdolon*, while *Anemone nemorosa* is rather constant. Furthermore *Fagus sylvatica* is increasing in both tree- and

Table 5. Development in a group of corresponding quadrats, selected on the basis of groups 5 and 10 in Table 2.

```
                        1935/35           1969/70           1976/76
                          1             1         1 1      11 11 1   11 111
Group number          55555555550424761  29225523528020267  29205002023002000
Field layer species:
ANEM RAN              12232111122-5-5-5  12--3221-323213-4  44-21324241434454
MERC PER              --2--12--2121-2-1  52-3-1331331213-3  544-11-3135-23--1
FRAX EXC              ----1-----1---1--  ---3111------41--  --12-1--1--11111-
ANEM NEM              454515443523541-3  52-5552145-4554-2  52-5435533-435121
LAMI GAL              43424443541-2-2--  -1--23223--------  -23-3------------
GAGE LUT              111------2--11---  ----------2------  -1---------------
GAGE MIN              -----------1-11--  -----------------  -----------------
GAGE SPA              11------2-11-----1  -1-----1---------  -1---------------
CHRY ALT              -----------2-1---  --1-------1----1--  --1--------------
CORY BUL              ------------2---1  ---1--------111-2  --------2--------
ADOX MOS              1-----------415---  -----------------  -3---------------
GEUM RIV              1-2---1--1-------  --2------2-------  -----------------
GEUM URB              11-------314-4---  --------1---1-1---  --1--------------
OXAL ACE              ---42----1-------  --------------- ---  --2-1------------
RANU FIC              -1221112121-1-354  -15-3-15-3^-2--51  -15----2124------
ST NE.GL              --4-1------------  ---1----- --------  -----------------
VICI SEP              -1--1----1-------  -----------------  -----------------

Tree and bush layer species:                1969               1975
ULMU GLA A1           -----5-------5---4  --21-41-------1--4  ----552--4--3---5
ULMU GLA A2           5--------------5  31----1---32----1  -54--4-4--445--33
ULMU GLA B1           -----------5-51--5  44---1-21-33---11  --4--14--545----5
ULMU GLA B2           ----------------3  --------1----11-1  -1313--214---41-1
FAGU SYL A1           ---55---5--------  ---542-453-35551-  ---2-4525--51554-
FRAX EXC A1           5---------5---5-5  12----3---52----3  44--2--5--4------
QUER ROB A1           5---------------  ------------1------  5--5---1----2----
CORY AVE B1           4-4---5--5553----  -4--------34---2-2  42------1------2--
FAGU SYL A2           -------5-4---5---  ----22-1----2----  ---545-3---------
FAGU SYL B1           -----5-5--45----  -----2-113-1323--  ---14441-5--513--
CRAT OXY B2           4-3----2---------  -1----11-----1---  -------1-1---1---
FRAX EXC B2           ----------------  -1----1----------  -113--------13---
Bottom layer species:  1935                                  1975
BRAC RUT              11-------1-----4-         ?             --1------2-----3-
EURH PRA              21----11-32---2-1         ?             ----------4------
EURH STR              -1---------------         ?             -4---------------
ATRI UND              -----11----------         ?             -----------------
MNIU UND              1--------1-------         ?             ----------1------
Characteristic Indicator Value, based on fiel  layer species only:
CONTINENTAL (K)       44434444444444433  34333444434333334  34334333333333344
LIGHT FIGURE(L)       44433333333434363  24533333334333353  23343433433333333
MOISTURE    (F)       55665666566666676  65666566567666676  66666666666666666
NITROGEN    (N)       66776666667777777  77777677676777768  77776777777888888
REACTION    (R)       77777776777676777  67766677677766777  77766767777777777
```

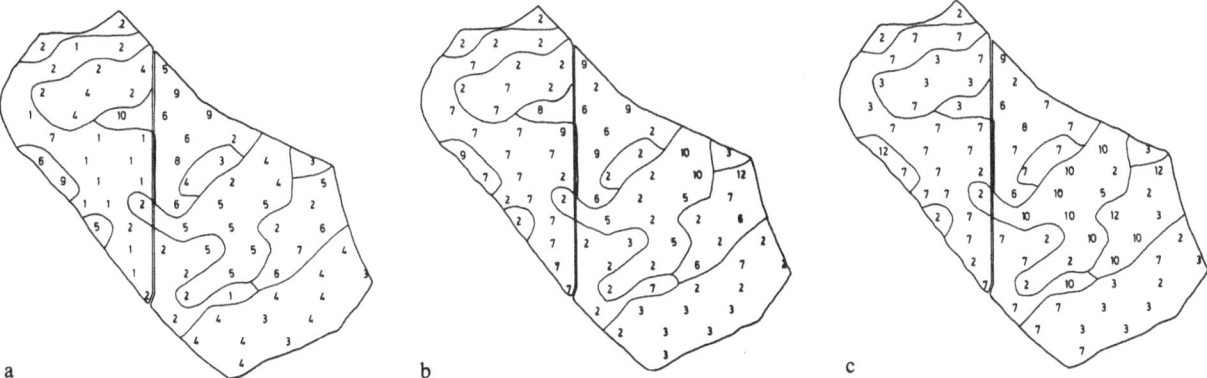

Fig. 7. Spatial distribution of vegetation types, and their changes in time. Each quadrat is represented by its group number from the classification. a) 1935/35 b) 1969/70 c) 1976/76.

shrub layers (A1 and B1), while *Fraxinus excelsior* is increasing in the B1 layer and as saplings in the field layer layer. Some trends are parallel to the trends in Table 4, e.g. the decreases for *Gagea lutea, G. minima, G. spathacea, Adoxa moschatellina, Geum rivale* and *Geum urbanum*. The increases for *Anemone ranunculoides* and *Mercurialis perennis* are also similar, but less pronounced. The increases for *Ulmus glabra* in the tree- and shrub layers are even more pronounced than in Table 4. The general decrease for mosses in the bottom layer are parallel to those in Table 4.

The characteristic indicator values for continentality and moisture in Tables 4 and 5 are rather constant in the course of time. The CIV's for light are decreasing, suggesting that average light conditions in these quadrats has changed from 'half' shadow to a somewhat deeper shadow (Ellenberg 1974). The values for nitrogen status show an increasing trend, indicating an increased importance of nitrogen indicators in the quadrats (value 8, Ellenberg op cit).

*Spatial-temporal patterns*

The spatial pattern of vegetation types and its changes in time were studied from maps, where each quadrat was represented by its group number from the classification (Fig 7). The area was divided into four subareas (I–IV, cf Fig 1) in order to represent the spatial-temporal pattern that emerges. The distribution of different groups at different times in these subareas (Table 6) demonstrates the strong relationship between the spatial variation and the temporal developments or successions already recognized in the ordination and the tabular analysis.

In subarea I the development is restricted to the upper part of the vegetational and successional gradient (Table 6). The subarea includes the majority of quadrats from groups 5 and 10 and thus represents the spatial element in the development illustrated in Table 5. The data for tree and shrub layers from the plots of this subarea indicate a development from a rather heterogeneous situation in 1935 towards an increased importance of *Fagus sylvatica* in layers A1, A2 and B1 (cf Table 5). Subarea II represents the same development as in Table 4, righthand part, as it includes all quadrats from group 1. These quadrats show a spatial concentration to the west, central part of the wood (Fig 7a). From this 'centre' the vegetation type dominated by *Anemone ranunculoides* and *Mercurialis perennis* (group 7) has developed progressively and attains a much wider distribution in the subsequent investigations (Fig 7a-c, Table 6).

Subarea III represents the spatial pattern of Table 4, left half, since it includes the majority of quadrats from group 4, located in the south end of the wood and in an area near the north end (Fig. 7a). The succession in this subarea has by 1976 proceeded 'one step' further along the gradient compared to subarea II (Table 6). This difference was indicated already in the classification, where the occurrences of *Filipendula ulmaria* and *Chrysosplenium alternifolium* in group 4 indicated a stronger link, compared to group 1, between group 4 and the far right end of Table 2. Subarea IV is a heterogeneous and rather constant area, with no clear trend of changes (Table 6). It contains the quadrats situated in wet depressions (groups 6 and 12), which changed rather little since 1935. It contains also the *Aegopodium podagraria* variant of the *Anemone nemorosa-A. ranunculoides* community (group 9), which tends to be located near the edges of the wood (Fig. 7a-c).

Table 6. The number of quadrats belonging to different groups in the four subareas in Figs 1 and 7. The table summarizes the spatial and temporal trends in the area. Sections I-IV are identical to those in Table 2.

| Group number | Area I (n=14) 35/35 | 69/70 | 76/76 | Area II (n=22) 35/35 | 69/70 | 76/76 | Area III (n=21) 35/35 | 69/70 | 76/76 | Area IV (n=17) 35/35 | 69/70 | 76/76 | Total (n=74) 35/35 | 69/70 | 76/76 | |
|---|---|---|---|---|---|---|---|---|---|---|---|---|---|---|---|---|
| 4 | 2 | - | - | - | - | - | 13 | - | - | - | - | - | 15 | - | - | Section I |
| 5 | 7 | 3 | 1 | 1 | - | - | - | - | - | 2 | - | - | 10 | 3 | 1 | Section I |
| 10 | - | 2 | 6 | - | - | 1 | - | - | - | 1 | - | 2 | 1 | 2 | 9 | |
| 1 | - | - | - | 13 | - | - | - | - | - | - | - | - | 13 | - | - | |
| 9 | - | - | - | - | 1 | - | - | - | - | 3 | 4 | 1 | 3 | 5 | 1 | |
| 2 | 5 | 9 | 7 | 7 | 7 | 3 | 3 | 11 | 2 | 3 | 4 | 2 | 18 | 31 | 14 | Section II |
| 7 | - | - | - | 1 | 13 | 17 | - | 3 | 8 | 1 | 2 | 4 | 2 | 18 | 29 | Section III |
| 3 | - | - | - | - | 1 | 1 | 5 | 7 | 11 | - | - | 2 | 5 | 8 | 14 | |
| 8 | - | - | - | - | - | - | - | - | - | 1 | 1 | 1 | 1 | 1 | 1 | |
| 6 | - | - | - | - | - | - | - | - | - | 6 | 5 | 2 | 6 | 5 | 2 | Section IV |
| 12 | - | - | - | - | - | - | - | - | - | - | 1 | 3 | - | 1 | 3 | |

*Vegetational and successional gradient* (vertical axis label spanning group numbers)

The intentionally misplaced sample, 1976/76-Alt, produces a spatial pattern very similar to the 1976/76 sample.

## Discussion

*Accuracy*

The problem of possibly deviating positions of quadrats, supposed to represent the same plot in the course of time, arises when comparisons are made between the three data matrices from 1935, 1969/70, and 1976 resp. The data can be considered as three separate, systematic samples of the entire wood. If the total variation and vegetation types encountered in each sample are considered, the position problem applies to the question whether the different samples are comparable. The consistency of the results in the ordinations (Fig. 3) and in the test with alternative, misplaced quadrats indicates that the samples are indeed comparable.

If comparisons between corresponding quadrats from different points in time are made, this should be done on a collective rather than an individual basis (cf Malmer et al 1978). This means that the same subset of quadrats is selected from each data matrix, forming what is here referred to as groups of corresponding quadrats. The data from such groups are interpreted under the assumption that effects of misplacements in individual quadrats are overruled by major trends in the data of the whole group. The results suggest that this is a reasonable assumption (e.g. Table 4, Figs 5 and 6).

However, a detailed comparison of individual corresponding quadrats is not possible with the present data, because of the effects of misplacements, but also because of other sources of sample errors such as year-to-year variation and personal variation in cover estimates between the different investigators.

The seriousness of deviating positions of quadrats, considered to represent the same sites at different times, must be judged in view of the considerable homogeneity of the present vegetation in large areas of the wood. In such areas a misplacement of a quadrat has a small effect on the results as was demonstrated in the test with alternative quadrats. Other areas like the wet depressions, which are both small in size and have sharper boundaries towards surrounding vegetation types, are more vulnerable to the effects of misplacement. Situations like these are responsible for the very low similarities between some corresponding quadrats in matrices 1976/76 and 1976/76-Alt (Fig 4, bottom). Such areas are on the other hand, easily recognizable in the field, and if measurements indicate a position outside such an area, the error is obvious and can be corrected.

*Methodological remarks*

Three different, though largely overlapping, classifications based on vegetation data from Dalby Söderskog

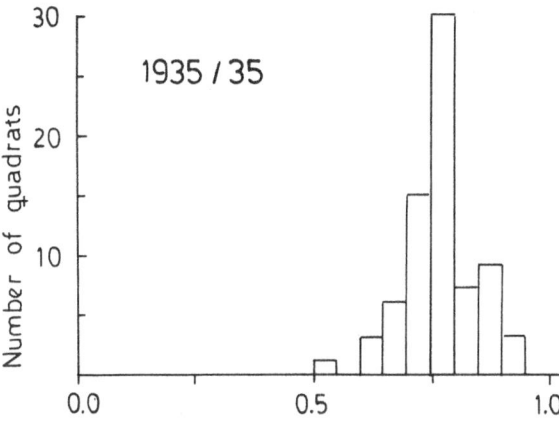

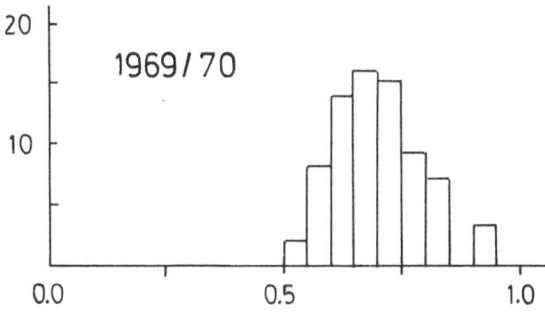

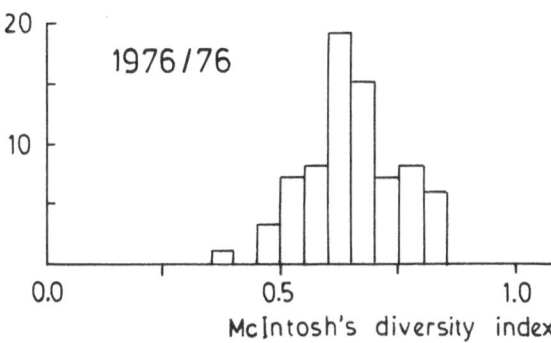

Fig. 8. The distribution of McIntosh's diversity index in the different samples. The diversity is calculated for each quadrat from the cover values for each species in a quadrat.

were presented earlier (Lindquist 1938, Lindgren 1971 and Malmer et al 1978). An analysis of changes based on any of these classifications would reveal the major trends, while minor trends may be focused differently by these not fully congruent classifications. This stresses the usefulness of a more objective numerical method, such as the TABORD-program. In Malmer et al (1978) four main oligothetic groups were distinguished, in the present paper eleven rather clear cut polythetic groups are recognized.

Three different ordinations were tested on the data, viz. centered, non-standardized principal component analysis (PCA), centered and standardized PCA with standardization of species to zero mean and unit variance and finally reciprocal averaging (Hill 1973). Of these the first method was most successful in giving a balanced picture of the variation in the data (cf Fig. 3). Standardization gave too much emphasis on rare species, especially reciprocal averaging was strongly effected by the occurrence of rare species.

*Plant successions*

The tree canopy and the shrub layers closed up in the wood, and both stem numbers and stem volume for trees with > 10 cm dbh increased between 1916 and 1970. A great part of this increase is due to an almost fourfold increase in the number of elms with > 10 cm dbh. The frequency of *Ulmus glabra* in the two shrub layers of the systematic sample also increased between 1935 and 1975, from 34 to 68 % in the B1 layer and from 11 to 46 % in the B2 layer (Malmer et al 1978). The majority of oaks are very old and virtually no regeneration of this species occurs. The oaks are overgrown by elms and ashes, thus suffering from both shadow and mechanical damage, as the slender stems of elms and ashes whip their crowns. On the basis of these observations it is reasonable to assume that the total amount of leaf litter in the wood has increased, and that the relative contributions from different tree species to this litter must have changed considerably. Thus the bulk of the present leaf litter consists of *Ulmus glabra* leaves, while the contribution from *Quercus robur* leaves has declined.

The response of the field layer species to these changes in the higher layers can be deduced from the trends seen in the selected corresponding quadrats, where rather drastic changes are indicated for several species. A large group of species show a general and marked decrease in frequency: *Adoxa moschatellina, Gagea minima, Corydalis fabacea, Chrysosplenium alternifolium, Lamium galeobdolon, Gagea spathacea, G. lutea* and *Geum urbanum*. A great deal of the changes observed are however due to changes in the relative quantities of some species, rather than changes in frequencies. Thus has the frequency of *Mercurialis perennis* been rather constant: 81, 93 and 88 % in 1935, 1969 and 1976 resp., but the species has increased its cover considerably. This is for example

119

illustrated in the converging succession towards a total dominance of *Mercurialis perennis* in the summer aspect of quadrats belonging to groups 7 and 3 (Figs 5 and 6, Table 2). The changes in frequency for *Anemone nemorosa* and *A. ranunculoides* are also moderate: 95, 92 and 84 % for *A. nemorosa*, 69, 91 and 92 % for *A. ranunculoides*, years 1935, 1970 and 1976 resp. A great deal of the changes observed in the spring aspect are however due to a changing balance between these two codominating species. This development is to a great extent parallel to the increasing dominance of *Mercurialis perennis* and is thus illustrated in the same converging succession as mentioned above.

The three samples show a gradual decline in diversity in the area during the observation period (Fig 8). This decline is partly due to a decrease in the number of species in each submatrix, from 58 species in 1935/35 to 35 species in 1969/70 and 36 species in 1976/76. The decline is also due to the development towards nearly total dominance of one species in many quadrats, so that quadrats with a number of species sharing the available space more or less equally are rarely encountered in the later samples. This type of divers quadrats are e.g. found in group 9 in the classification (Table 2).

There is no single environmental gradient to be paralleled with the vegetational and successional gradients recognized. A response to the closer canopy can be traced in the decreasing CIV's for light, indicating that shadow tolerant species has become more important in the quadrats. A development towards a richer habitat is indicated in many plots by the increase of demanding species, i.e. species normally associated with high levels of available nutrients and basic to circumneutral soil reaction, notably *Anemone ranunculoides* and *Mercurialis perennis*, at the expense of less demanding species, e.g. *Anemona nemorosa* and *Lamium galeobdolon* (Table 4 and 5). This is also reflected in the increases in CIV's for nitrogen status.

The decomposition of litter from different sources was investigated by Lindquist. He found that equal amounts of elm and oak leaves were deposited on the 74 sq. meters, covered by the systematic plot system, in November 1935 (5556 and 5206 grams dry weight resp.). Making these amounts equal to 100 %, he found that by May 1936 only 3 % of the elm litter remained as dry weight, while 82 % of the oak leaves remained. The corresponding figs. for August 1936 were 1 and 32 % (Lindquist 1938). Earthworms and other macrodecomposers attacked oak- and beech litter only after the supply of elm-, ash-, hazel-

and *Mercurialis* litter was consumed (Lindquist op cit). The acidity of litter from different sources was found to vary from *Mercurialis perennis* litter with pH = 7.0, over *Ulmus glabra* pH = 6.9, *Fraxinus excelsior* pH = 6.8, *Corylus avellana* pH = 6.5, *Fagus sylvatica* pH = 6.2 down to *Quercus robur* litter with pH = 5.5 (Lindquist 1938). Values for ash content (%) in litters show almost the same descending series: *Mercurialis* 16.9, *Ulmus* 21.3, *Fraxinus* 15.3, *Corylus* 11.4, *Fagus* 6.7 and *Quercus* 7.5 (Lindquist op cit).

*Ulmus glabra* leaves thus give a richer litter with lower acidity, higher ash content and faster decomposition rate, as compared to *Quercus* leaves. These differences, coupled with the assumed quantitative changes in total litter and its composition, may well provide the richer habitat mentioned above.

A somewhat restricted development, with almost constant cover and frequency for *Anemone nemorosa* and less pronounced increases for *A. ranunculoides* and *Mercurialis perennis*, was observed in subarea I, where *Fagus sylvatica* is abundant in tree- and bush layers (cf Table 5 and 6). This subarea has a slower decomposition rate of litter, so that a layer of undecomposed *Fagus* leaves persists in August. The general improvement of the habitat through the increased *Ulmus* litter is therefore less pronounced in this subarea, which may account for the restricted development in the fieldlayer.

## Summary

Successions in a South Swedish deciduous wood, which is a national park and has been almost undisturbed since 1918, are studied using semipermanent plots. These plots were originally established in 1935 and reestablished in 1969–70 and 1975–76. Each plot is represented by a record, which combines fieldlayer data from spring- and summer aspects. The accuracy of the approach of semipermanent plots is evaluated from a slightly displaced sample.

Vegetational and successional gradients are revealed in a classification of the entire dataset. Successional trends are analysed through ordinations of site- period data. A converging successional trend is found. The vegetational and successional gradients are found to coincide. A considerable reduction in the variation and diversity of the fieldlayer is evident.

An analysis of the development in smaller groups of corresponding quadrats shows consistent, and in some cases rather drastic, changes in cover and/or frequency for many species. Characteristic indicator values are used to connect these changes to environmental factors, and changes in light regime and nitrogen status are indicated. The spatial-temporal pattern is analysed through maps.

## Appendix

Species codes and species names of Tables 2, 4 and 5. (species belonging to the spring aspect are marked*)

| | | |
|---|---|---|
| ACTA | SPI | Actaean spicata |
| ADOX | MOS | Adoxa moschatellina* |
| AEGO | POD | Aegopodium podagraria |
| ANEM | NEM | Anemone nemorosa* |
| ANEM | RAN | Anemone ranunculoides* |
| ANGE | SIL | Angelica silvestris |
| ANTH | SIL | Anthriscus silvestris |
| C | REMOTA | Carex remota |
| C | SILVAT | Carex silvatica |
| CALT | PAL | Caltha palustris* |
| CAMP | LAT | Campanula latifolia |
| CAMP | TRA | Campanula trachelium |
| CHRY | ALT | Chrysosplenium alternifolium* |
| CIRC | LUT | Circaea lutetiana |
| CORY | BUL | Corydalis bulbosa* |
| CORY | FAB | Corydalis fabacea* |
| CORY | AVE | Corylus avellana |
| CRAT | OXY | Crataegus oxyacantha |
| CREP | PAL | Crepis paludosa |
| DACT | GLO | Dactylis glomerata |
| EPIL | MON | Epilobium montanum |
| EPIP | HEL | Epipactis helleborine |
| EQUI | SIL | Equisetum silvaticum |
| FAGU | SYL | Fagus sylvatica |
| FEST | GIG | Festuca gigantea |
| FILI | ULM | Filipendula ulmaria |
| FRAG | VES | Fragaria vesca |
| FRAX | EXC | Fraxinus excelsior |
| GAGE | LUT | Gagea lutea* |
| GAGE | MIN | Gagean minima* |
| GAGE | SPA | Gagea spathacea* |
| GALI | APA | Galium aparine |
| GERA | ROB | Geranium robertianum |
| GEUM | RIV | Geum rivale |
| GEUM | URB | Geum urbanum |
| LAMI | GAL | Lamium galeobdolon |
| LATH | VER | Lathyrus vernus* |
| LONI | XYL | Lonicera xylosteum |
| MERC | PER | Mercurialis perennis |
| OXAL | ACE | Oxalis acetosella |
| PARI | QUA | Paris quadrifolia |
| POA | NEM | Poa nemoralis |
| POA | TRI | Poa trivialis |
| POLY | MUL | Polygonatum multiflorum |
| PULM | OFF | Pulmonaria officinalis* |
| QUER | ROB | Quercus robur |
| RANU | FIC | Ranunculus ficaria* |
| RANU | REP | Ranunculus repens |
| RUME | SAN | Rumex sanguineus |
| SANI | EUR | Sanicula europaea |
| ST | NE.GL | Stellaria nemorum ssp glochidisperma |
| STAC | SIL | Stachys silvatica |
| TARAXACZ | | Taraxacum sp |
| ULMU | GLA | Ulmus glabra |
| URTI | DIO | Urtica dioica |
| VERO | CHA | Veronica chamaedrys |
| VERO | HED | Veronica hederifolia |
| VERO | MON | Veronica montana |
| VIBU | OPU | Viburnum opulus |
| VICI | SEP | Vicia sepium |
| VIOLAZ | | Viola sp |

## References

Austin, M.P. 1977. Use of ordination and other multivariate descriptive methods to study succession. Vegetatio 35: 165–175.

Ellenberg, H. 1974. Zeigerwerte der Gefässpflanzen Mitteleuropas. Scripta Gebot 9. Göttingen, 97 pp.

Hill, M.O. 1973. Reciprocal averaging: an eigenvector method of ordination. J. Ecol. 61: 237–249.

Janssen, J.G.M. 1972. Detection of some micropatterns of winter annuals in pioneer communities of dry sandy soils. Acta Bot. Neerl. 21: 609–616.

Lance, G.N. & W.T. Williams. 1968. Note on a new information statistic classification program. Comput. J. 11: 195.

Lid, J. 1974. Norsk og svensk flora. Oslo, 808 pp.

Lindgren, L. 1971. Skötsel av lövskogsområden. Vegetationsförändringar i Dalby Söderskog. Meddelanden från forskargruppen för skötsel av naturreservat nr. 11. 43 pp + 12 app.

Lindquist, B. 1938. Dalby Söderskog, en skånsk lövskog i forntid och nutid. Acta Phytogeogr. Suec. 10, 273 pp. (summary in German).

Maarel, E. van der. 1969. On the use of ordination models in phytosociology. Vegetatio 19: 21–46.

Maarel, E. van der, J.G.M. Janssen & J.M.W. Louppen. 1978. TABORD, a program for structuring phytosociological tables. Vegetatio 38: 143–156.

Malmer, N., L. Lindgren & S. Persson. 1978. Vegetational succession in a south Swedish deciduous wood. Vegetatio 36: 17–29.

McIntosh, R.P. 1967. An index of diversity and the relation of certain concepts to diversity. Ecology 48: 392–404.

Nyholm, E. 1954–69. Illustrated moss flora of Fennoscandia. II. Lund, 799 pp.

Persson, S. 1977. Datorprogram för bearbetning av vegetationsdata. 1. Klassifikationsprogram – dokumentation och handhavande. Meddn Avd Ekol. Bot., Lunds Univ. 33. 68 pp.

Roskam, E. 1971. Program ORDINA: Multidimensional ordination of observation vectors. Programme-Bulletin 16. Dept. of Psychology, University of Nijmegen, 8 pp.

Trass, H. & N. Malmer. 1978. North European approaches to

classification. 2nd. ed. In: R.H. Whittaker: Classification of plant communities, p. 207–245. Junk, The Hague.

Westhoff, V. & E. van der Maarel. 1978. The Braun-Blanquet approach. 2nd. ed. In: R.H. Whittaker (ed.), Classification of plant communities, p. 287–399. Junk, The Hague.

Williams, W.T., G.N. Lance, L.J. Webb, J.G. Tracey & M.B. Dale. 1969. Studies in the numerical analysis of complex rainforest communities. III. The analysis of successional data. J. Ecol. 57: 515–536.

Wishart, D. 1969. CLUSTAN Ia. User manual. St Andrews Computing Centre. St Andrews.

Österdahl, L., G. Zetterberg & I. Andersson. 1977. Introduktion till RUBIN. SNV PM 909. 148 pp + 4 app.

Accepted 14 November 1979

# A NUMERICAL STUDY OF SUCCESSIONS IN AN ABANDONED, DAMP CALCAREOUS MEADOW IN S SWEDEN*,**

G. REGNÉLL***

Department of Plant Ecology, University of Lund  Östra Vallgatan 14, S-223 61 Lund, Sweden

**Keywords:**
*Caricion davallianae*, *Filipendula ulmaria*, Nature conservation, Pasture, PCA-ordination, Succession, Sweden, Tussock

## Introduction

Vegetation corresponding to the *Caricion davallianae* (Westhoff & den Held 1969, Klika 1934) is very rare in S Sweden. It is confined to calcareous soils (calcareous moraine, shell deposits, calcareous tufa etc) with a fairly high or moderately fluctuating subsoil water table. (See Regnéll 1976 for a literature review).

This kind of vegetation, rich in species, many of them rare, was formerly widespread, but it requires active conservation to preserve it. The traditional hay-making or grazing will soon have stopped, and with these practices this vegetation will disappear. To find effective and economical management measures, a better knowledge of the vegetation is needed, especially regarding the succession of the different stages; the vegetation dynamics.

No permanent plots of this vegetation type exist in Sweden, so an indirect approach was tried, viz. a comparison between a grazed area and an adjacent area which was not used for grazing for ca. 15 years, but according to local people, used to have a very similar vegetation.

The studied area is near Örup, 13 km north-east of the port of Ystad in southernmost Sweden. Here along the Örupsån (Örup river), vast unimproved meadows are situated. For centuries they were used for haymaking (Geometriska kartan 1727, filed by the county administration in Kristianstad), but during the twentieth century they

* Nomenclature of vascular plants follows Lid (1974), for names of bryophytes see Nyholm (1954-69) and Arnell (1956).
** This paper is mainly based on a more detailed paper in Swedish (Regnéll 1979).
*** I thank Dr. E. van der Maarel, Prof. Nils Malmer, Fil. lic. Anders Larsson, Fil. Dr. Eva Waldemarsson-Jensén and Fil. kand. Stefan Persson for valuable discussions and encouragement.

have only been used for grazing. The underlying moraine is rich in calcareous clay but it also contains large archaean boulders.

## Methods

### Field observations

Two plots, A and C, each 100 m², were chosen in a vegetation as homogeneous as possible. The area where plots A and C are situated is grazed with an average yearly pressure of 50-100 cowdays/ha. There is, however, a variation within the area, the grazing in A being satisfactory from a nature conservancy point of view ('heavy') while C is less intensely grazed. Within each plot 20 0.1 m² quadrats were distributed at random, though each was displaced as much as 40 cm in a direction settled beforehand if this was necessary (and sufficient) to get the quadrat on a distinct tussock or in a distinct depression. The displacement was made in order to make the differences between these two elements in the vegetation more clearly distinguishable.

On the other side of a fence the vegetation had not been grazed since ca. 1960 and the variation was less evident (no tussocks, fewer species). Here, 10 quadrats were distributed in plot B. An experiment with different management regimes also started here in three adjacent 9 m² plots, each with 10 0.1 m² quadrats. In this paper these 30 quadrats are treated together as 'plot' D. Plot B was wetter than D and inundated for a longer time in the early spring, like some of the depressions between the tussocks in plots A and C. However, the water conditions in plots A and C are on the whole the same as in plot D.

In each 0.1 m² quadrat all the species present were

determined and their percentage cover estimated. Notes were also made regarding the height of the vegetation, the percentage cover of litter, and whether the quadrat was situated on a tussock or in a depression (in relation to the immediate surroundings). The highest point in every quadrat was related to a local benchmark by levelling.

Observations were made regarding the positions (tussock or depression) in which freely grazing cattle placed their hooves.

### Numerical methods

The 80 quadrats were ordinated by PCA (principal component analysis, unstandardized) using the program ORDINA (Roskam 1971) with % cover values for all 95 species present (Fig. 1). A study was also made regarding the effect of transformation into the essentially logarithmic Hult-Sernander-Du Rietz scale (Du Rietz 1921), from now on called 'HULT' (Fig. 2). This scale, traditionally used in Sweden, has five degrees of cover: 5: $> \frac{1}{2}$, 4: $\frac{1}{2}-\frac{1}{4}$, 3: $\frac{1}{4}-\frac{1}{8}$, 2: $\frac{1}{8}-\frac{1}{16}$, 1: $< \frac{1}{16}$.

The program DISPLAY (Persson 1978) was used as a tool for the interpretation of the ordination results. With this program numerical values can be printed on the positions of the quadrats in the ordinations. The numerical values may concern either data used in the ordination (i.e. cover values of a certain species, Figs 3-4), or additional independent information, e.g. about the height of the vegetation, litter cover, level (vertical position above the benchmark) and number of species (Figs 7-10). If a correlation between the ordination results and such independent data appears, a causal or functional connection may be supposed.

To summarize the information on the species composition of the four plots A-D two measures were calculated for the four groups of quadrats: the frequency, $F$, and the characteristic degree of cover, $C$, i.e. the mean cover % calculated over the quadrats in which a species occurs (Malmer 1962, p. 49).

### Results

#### Ordinations based on % and HULT cover values

Ordination results based on % cover values are very much affected by a minority of species strongly dominating some quadrats (Fig 1). Thus, the cover values of *Filipendula ulmaria* and *Carex disticha* (Figs 3-4) in the quadrats are the main determinants of the ordination pattern in Fig. 1.

Table 1. Main differences between vegetation of grazed (A, C) and ungrazed (B, D) plots. – The complete species composition of the plots is given in Regnéll (1979).

| Plot | Grazed | | Ungrazed | |
|---|---|---|---|---|
| | A | C | B | D |
| No. of quadrats/plot | 20 | 20 | 10 | 30 |
| Tot. no. of species/plot | 70 | 66 | 16 | 11 |
| Mean no. of species/quadrat | 15,5 | 17,0 | 5,5 | 5,5 |
| Mean height of veg. (cm) | 15,7 | 34,2 | 50,0 | 54,5 |

| | F | C | F | C | F | C | F | C |
|---|---|---|---|---|---|---|---|---|
| *Carex panicea* | 95 | 14 | 85 | 19 | . | . | . | . |
| *Festuca ovina* | 40 | 32 | 30 | 19 | . | . | . | . |
| *Molinia caerulea* | 80 | 22 | 65 | 11 | . | . | . | . |
| *Centaurea jacea* | 50 | 7 | 10 | 12 | . | . | . | . |
| *Campylium stellatum* | 50 | 11 | 15 | 2 | 10 | 1 | . | . |
| *Carex flacca* | 30 | 5 | 45 | 4 | . | . | . | . |
| *Galium uliginosum* | 10 | 2 | 45 | 3 | . | . | . | . |
| *Potentilla erecta* | 50 | 12 | 60 | 4 | . | . | . | . |
| *Serratula tinctoria* | 50 | 6 | 50 | 61 | . | ` | ` | . |
| *Thuidium tamariscinum* | 25 | 15 | 50 | 16 | . | . | . | . |
| *Carex nigra* | 65 | 10 | 75 | 6 | 80 | 13 | . | . |
| *Filipendula ulmaria* | 10 | 6 | 75 | 47 | 30 | 7 | 90 | 83 |
| *Carex disticha* | 45 | 9 | 90 | 4 | 100 | 67 | 100 | 24 |
| *Urtica dioica* | . | . | . | . | . | . | 93 | 17 |
| *Galium aparine* | . | . | . | . | . | . | 80 | 2 |

The variance accounted for is therefore high; 46, 19 and 7 % on the first three axes.

The ordination clearly separates ungrazed quadrats of plots B and D from heavily grazed ones (plot A) with some of the lightly grazed ones (plot C) in between (Fig. 1). The wetter quadrats of plot B are mainly dominated by *Carex disticha* and separated from the others. Tussocks and depressions are only distinguishable on the grazed side; they occur also separated in the ordination, but this is much more evident along the third axis (not shown here).

Transformation into the (logarithmic) HULT scale weakens the influence of the quantitatively dominating species. Only 29% and 15% of the variance is accounted for by axes 1 and 2 (Fig. 2). Nevertheless, the ecological interpretation is clearer (cf Jensén 1978): the essential information is now concentrated in the diagram of axes 1 and 2. The ungrazed quadrats, all strongly dominated by combinations of the same one to three species, appear clustered in Fig. 2. The grazed ones, having a greater species variation, are scattered in significant groups. Tussocks and depressions are distinguishable. Species important for this differentiation are e.g. *Carex panicea* and *Festuca ovina* (Regnéll 1979). However, the cover of *Filipendula ulmaria* and *Carex disticha* (Figs 5-6) still is the main factor determining the result of the ordination.

On the whole the tussocks accomodate more species than the depressions. In total, 45 species were found in

124

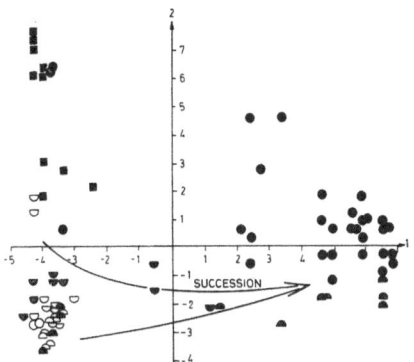

Fig. 1. Ordination by PCA, using percent cover. Axes 1 and 2. Symbols denote grazing intensity in the plots and the microtopographical position of the quadrats.

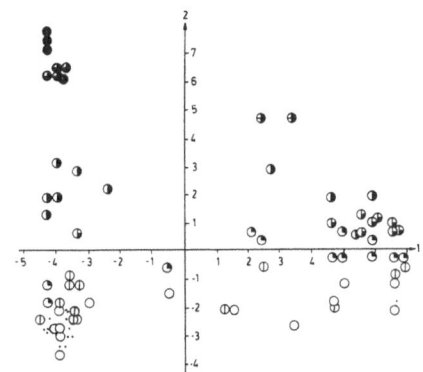

Fig. 4. Cover of *Carex disticha*, plotted on the PCA quadrat ordination of Fig. 1.

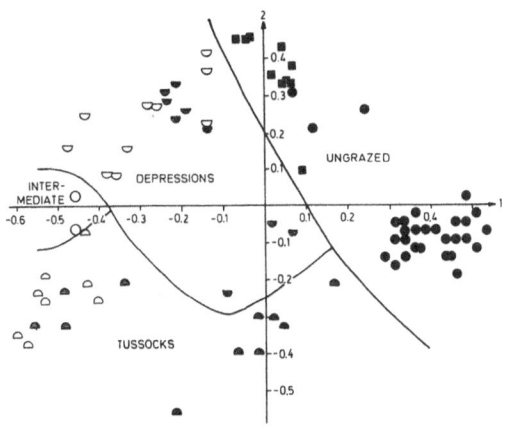

Fig. 2. Ordination by PCA, using HULT transformation of cover. Axes 1 and 2. Symbols as in Fig. 1.

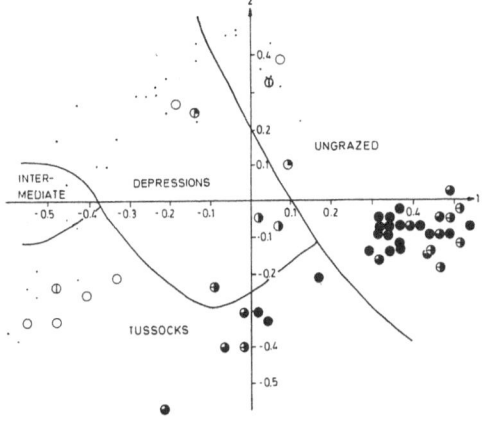

Fig. 5. Cover of *Filipendula ulmaria*, plotted on the PCA quadrat ordination of Fig. 2.

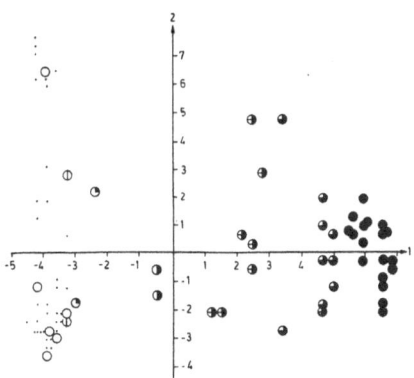

Fig. 3. Cover of *Filipendula ulmaria*, plotted on the PCA quadrat ordination of Fig. 1.

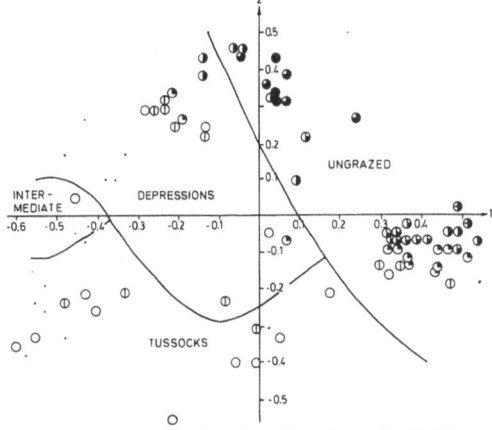

Fig. 6. Cover of *Carex disticha*, plotted on the PCA quadrat ordination of Fig. 3.

| Symbols of Figs 3-6 | · | ○ | ◐ | ◑ | ◒ | ◓ | ◕ | ◕ | ● |
|---|---|---|---|---|---|---|---|---|---|
| Cover classes (%) | 0 | 1,2 | 3,5 | 10,15 | 20,25 | 30,40 | 50,60 | 70,80 | 90– |

125

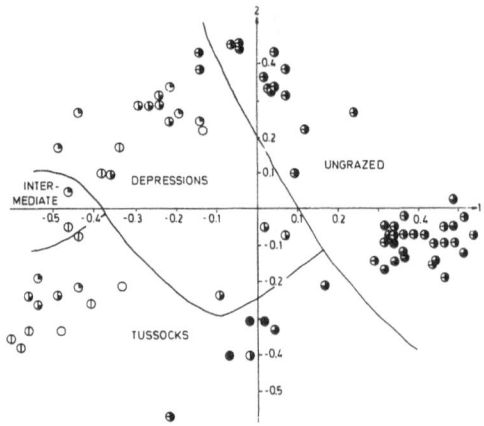

Fig. 7. Height of vegetation in the quadrats, plotted on the PCA quadrat ordination of Fig. 2. Symbols denote height in cm. cm.

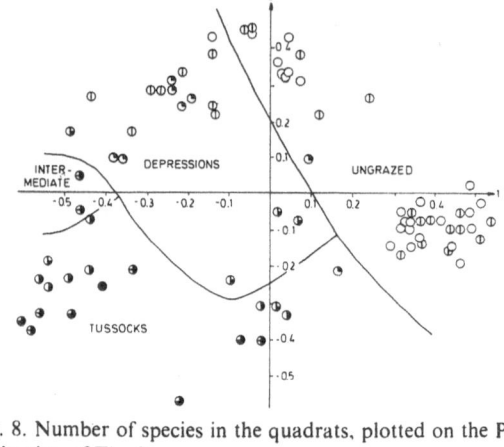

Fig. 8. Number of species in the quadrats, plotted on the PCA ordination of Fig. 2.

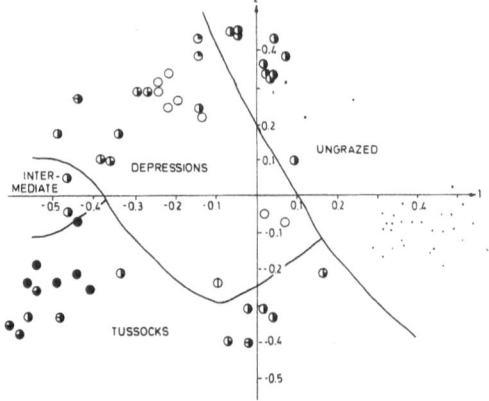

Fig. 9. Level of the quadrats in relation to the lowest situated depression, plotted on the PCA ordination of Fig. 2.

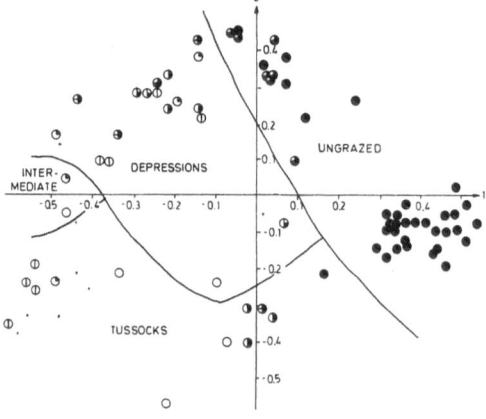

Fig. 10. Litter cover in the quadrats, plotted on the PCA ordination of Fig. 2.

| | | | ○ | ◑ | ◐ | ◑ | ◐ | ◕ | ◕ | ● |
|---|---|---|---|---|---|---|---|---|---|---|
| Fig. 7 | Veg. height (cm) | | < 6 | 6 - 10 | 11 - 15 | 16 - 25 | 26 - 35 | 36 - 55 | 56 - 75 | > 75 |
| Fig. 8 | Species number | | < 6 | 6 - 9 | 10 - 13 | 14 - 17 | 18 - 21 | 22 - 25 | 26 - 29 | > 29 |
| Fig. 9 | Level (cm) | * | < 6 | 6 - 10 | 11 - 15 | 16 - 20 | 21 - 25 | 26 - 30 | 31 - 35 | > 35 |
| Fig.10 | Litter cover (%) | 0 | 1 - 5 | 6 - 10 | 11 - 15 | 16 - 25 | 26 - 40 | 41 - 60 | 61 -.80 | 81 - 100 |

* A dot in Fig. 9 denotes that levelling was not made.

depressions, 78 on the tussocks. This is also true for individual quadrats: mean numbers of species are 11.4 and 8.9 in the depressions and 21.5 and 22.2 on the tussocks in plots A and C, respectively (Table 2, Fig. 8).

The height of the vegetation (Fig. 7) is clearly related to the changes in species composition when grazing ceases.

This is also true for litter accumulation (Fig. 10). The ungrazed area has almost a 100% cover of litter under the tall herbs. Bryophytes are scarce, growing on the litter without real contact with the soil surface. Fruits and seeds do not reach the ground easily.

126

Table 2. A comparison between the vegetation of 19 quadrats in depressions and 19 quadrats of tussocks in plots A and C.

| Plot | Depressions C | | Depressions A | | Tussocks C | | Tussocks A | |
|---|---|---|---|---|---|---|---|---|
| No of quadrats | 9 | | 10 | | 11 | | 8 | |
| Mean no of sp/quadrat | 11,4 | | 8,9 | | 21,5 | | 22,2 | |
| Total no of species | 36 | | 28 | | 55 | | 60 | |
| Total no of species | 45 | | | | 78 | | | |
| | F | C | F | C | F | C | F | C |
| Galium palustre | 44 | 2 | . | . | . | . | . | . |
| Caltha palustris | 56 | 10 | 20 | 8 | . | . | . | . |
| Deschampsia caespitosa | 33 | 2 | 20 | 4 | . | . | . | . |
| Galium uliginosum | 67 | 3 | 20 | 2 | 27 | 1 | . | . |
| Carex flacca | 33 | 5 | 50 | 5 | 55 | 4 | . | . |
| Selinum carvifolia | 44 | 6 | . | . | 100 | 7 | . | . |
| Carex hirta | 11 | 1 | . | . | 36 | 2 | . | . |
| C. hostiana | 33 | 3 | 70 | 16 | . | . | 25 | 1 |
| C. disticha | 100 | 6 | 50 | 15 | 82 | 2 | 38 | 2 |
| C. nigra | 89 | 8 | 80 | 15 | 64 | 4 | 38 | 1 |
| Calliergonella cuspidata | 78 | 6 | 60 | 13 | 18 | 1 | 25 | 1 |
| Carex panicea | 89 | 32 | 90 | 25 | 82 | 6 | 100 | 4 |
| Campylium stellatum | 22 | 3 | 50 | 19 | 9 | 1 | 38 | 5 |
| Filipendula ulmaria | 44 | 28 | 10 | 10 | 100 | 58 | 12 | 1 |
| Festuca rubra | 11 | 1 | 30 | 1 | 55 | 2 | 25 | 2 |
| Molinia caerulea | 56 | 7 | 60 | 25 | 73 | 13 | 100 | 15 |
| Salix repens | 11 | 2 | 10 | 15 | 22 | 10 | 38 | 37 |
| Lathyrus pratensis | 11 | 1 | 10 | 1 | 45 | 2 | 50 | 2 |
| Eriophorum angustifolium | . | . | 70 | 1 | . | . | 12 | 1 |
| Carex | 33 | 3 | . | . | 27 | 1 | 12 | 1 |
| Equisetum arvense | 56 | 1 | . | . | 55 | 4 | 12 | 1 |
| Cirsium palustre | 11 | 3 | . | . | 45 | 3 | 12 | 1 |
| Fissidens adianthoides | 11 | 1 | . | . | 36 | 1 | 50 | 1 |
| Geum rivale | 11 | 30 | . | . | 82 | 14 | 50 | 8 |
| Thuidium tamariscinum | 22 | 2 | . | . | 73 | 20 | 62 | 15 |
| Galium boreale | 22 | 1 | . | . | 91 | 5 | 88 | 2 |
| Succisa pratensis | 11 | 10 | . | . | 64 | 57 | 50 | 8 |
| Potentilla erecta | 11 | 1 | . | . | 100 | 4 | 100 | 14 |
| Serratula tinctoria | 11 | 1 | . | . | 82 | 7 | 100 | 7 |
| Ranunculus acris | . | . | 30 | 1 | 9 | 1 | 75 | 1 |
| Briza media | . | . | 20 | 2 | 18 | 2 | 62 | 2 |
| Centaurea jacea | . | . | 20 | 6 | 18 | 2 | 75 | 8 |
| Vicia cracca | . | . | 10 | 1 | 27 | 1 | 38 | 1 |
| Carex pulicaris | . | . | 10 | 1 | 36 | 1 | 38 | 1 |
| Sieglingia decumbens | . | . | 10 | 1 | 55 | 2 | 62 | 5 |
| Luzula multiflora | . | . | . | . | 27 | 1 | . | . |
| Rumex acetosa | . | . | . | . | 27 | 3 | . | . |
| Brachythecium curtum | . | . | . | . | 36 | 4 | . | . |
| Scorzonera humilis | . | . | . | . | 45 | 3 | 25 | 2 |
| Filipendula vulgaris | . | . | . | . | 45 | 1 | 38 | 3 |
| Climacium dendroides | . | . | . | . | 36 | 4 | 25 | 1 |
| Mnium affine | . | . | . | . | 36 | 1 | 25 | 1 |
| Anthoxanthum odoratum | . | . | . | . | 27 | 3 | 25 | 1 |

| Plot | Depressions C | | Depressions A | | Tussocks C | | Tussocks A | |
|---|---|---|---|---|---|---|---|---|
| | F | C | F | C | F | C | F | C |
| Dicranum scoparium | . | . | . | . | 36 | 31 | 62 | 15 |
| Festuca ovina | . | . | . | . | 55 | 19 | 88 | 35 |
| Valeriana dioica | . | . | . | . | 27 | 10 | 75 | 2 |
| Aulacomnium palustre | . | . | . | . | 18 | 20 | 75 | 19 |
| Leontodon hispidus | . | . | . | . | . | . | 38 | 5 |
| Prunella vulgaris | . | . | . | . | . | . | 38 | 2 |

Less frequent species

| | F | C | F | C | F | C | F | C |
|---|---|---|---|---|---|---|---|---|
| Brachythecium populeum | 11 | 2 | . | . | . | . | . | . |
| Trifolium pratense | 11 | 1 | . | . | . | . | . | . |
| Ranunculus flammula | 22 | 1 | 10 | 1 | . | . | . | . |
| Campylium elodes | 11 | 1 | 20 | 1 | . | . | . | . |
| Drepanocladus | . | . | 20 | 13 | . | . | . | . |
| Lysimachia vulgaris | 11 | 1 | . | . | 9 | 5 | . | . |
| Mentha arvensis | 22 | 1 | 10 | 1 | . | . | 12 | 1 |
| Drepanocladus revolvens | 11 | 1 | 20 | 10 | . | . | 12 | 1 |
| Ctenidium molluscum | 11 | 3 | . | . | . | . | 12 | 1 |
| Agrostis stolonifera | . | . | 10 | 1 | . | . | 12 | 1 |
| Salix aurita | . | . | . | . | 18 | 20 | . | . |
| Ranunculus auricomus | . | . | . | . | 18 | 2 | . | . |
| Viola canina | . | . | . | . | 18 | 2 | . | . |
| Lophocolea heterophylla | . | . | . | . | 18 | 1 | . | . |
| Crataegus | . | . | . | . | 9 | 60 | . | . |
| Rhytidiadelphus loreus | . | . | . | . | 9 | 2 | . | . |
| Arrhenatherum pubescens | . | . | . | . | 9 | 1 | . | . |
| Galium verum | . | . | . | . | 9 | 1 | . | . |
| Bryum-art | . | . | . | . | 9 | 1 | . | . |
| Mnium undulatum | . | . | . | . | 9 | 1 | . | . |
| Bryum pseudotriquetrum + bimum | . | . | . | . | 9 | 1 | 12 | 1 |
| Anemone nemorosa | . | . | . | . | 9 | 2 | 25 | 2 |
| Plantago lanceolata | . | . | . | . | . | . | 25 | 2 |
| Festuca arundinacea | . | . | . | . | . | . | 25 | 1 |
| Scleropodium purum | . | . | . | . | . | . | 25 | 1 |
| Hieracium auricula | . | . | . | . | . | . | 12 | 15 |
| Thuidium recognitum | . | . | . | . | . | . | 12 | 15 |
| Quercus robur (juv) | . | . | . | . | . | . | 12 | 5 |
| Lotus corniculatus | . | . | . | . | . | . | 12 | 2 |
| Arrhenatherum pratensis | . | . | . | . | . | . | 12 | 1 |
| Luzula campestris | . | . | . | . | . | . | 12 | 1 |
| Dactylorhiza· | . | . | . | . | . | . | 12 | 1 |
| Equisetum palustre | . | . | . | . | . | . | 12 | 1 |
| Linum catharticum | . | . | . | . | . | . | 12 | 1 |
| Ononis hircina | . | . | . | . | . | . | 12 | 1 |
| Leucobryum glaucum | . | . | . | . | . | . | 12 | 1 |
| Cladonia furcata | . | . | . | . | . | . | 12 | 1 |

## Vegetation of the tussocks and depressions

The tussocks are rich in species (Fig. 8); many of them are important in communities of dry habitats. E.g. *Arrhenatherum pratensis*, *A. pubescens*, *Leontodon hispidus*, and *Festuca ovina* all occur in *Festuco-Brometea* and *Filipendula vulgaris* in *Trifolio-Geranietea sanguinei* communities (Westhoff & den Held 1969). However the dominants are species with a wider ecological range such as *Molinia caerulea*, *Potentilla erecta*, *Succisa pratensis*, and, where grazing is light, *Filipendula ulmaria* and *Geum rivale*. Important bryophytes are *Fissidens adianthoides*, *Thuidium tamariscinum*, *Dicranum scoparium* and *Aulacomnium palustre*. Even the lightly grazed tussocks are rich in species. Here the vegetation is higher (Fig. 7) but the initial stages of change are not revealed by the number of species.

The depressions are characterized by *Caltha palustris*, *Galium palustre* and *G. uliginosum* (Table 2). Typical but less frequent are *Eriophorum angustifolium*, *Ranunculus flammula*, *Mentha arvensis* and *Drepanocladus spp*. Prevailing species are usually *Carex panicea*, *C. disticha*, *C. hostiana*, *C. nigra*, *Molinia caerulea* and *Filipendula ulmaria*, the latter though being even more common on the tussocks. *Campylium stellatum* and *Calliergonella cuspidata* are the most common bryophytes.

Table 3. Succession on the tussocks. To the left, grazed vegetation of the tussocks in plot A, to the right, ungrazed vegetation in plot D (not differentiated in tussocks and depressions). In between, quadrats from tussocks in plot C are placed along the succession line suggested in Fig. 1. Species occurring in one quadrat only in plot C or plots A or D only are excluded (given in Regnéll 1979).

| Tussocks in plot | A | C | | | | | | | | | | | D |
|---|---|---|---|---|---|---|---|---|---|---|---|---|---|
| Number of quadrats | | 41 | 43 | 42 | 46 | 34 | 39 | 31 | 36 | 40 | 45 | 38 | |
| Height of veg (cm) | m=11,8 | 10 | 5 | 25 | 3 | 30 | 40 | 80 | 95 | 70 | 85 | 75 | m=54,5 |
| Number of species | m=22,2 | 24 | 27 | 20 | 23 | 25 | 26 | 28 | 18 | 18 | 15 | 13 | m= 5,5 |
| Altitude (cm) | m=36,1 | 23 | 26 | 22 | 24 | 26 | 25 | 19 | 23 | 22 | 22 | 16 | |
| Cover of litter (%) | m= 5 | . | . | 15 | 5 | 30 | 5 | 5 | 40 | 40 | 50 | 100 | 100 |
| Depth of litter (cm) | m= 3 | . | . | 10 | 1 | 1 | 1 | 1 | 10 | 10 | 7 | 5 | 5 |
| | F–C | Cover (%) | | | | | | | | | | | F–C |
| Briza media | 62– 2 | 1 | 2 | . | . | . | . | . | . | . | . | . | . |
| Luzula multiflora | . | 1 | 1 | . | 2 | . | . | . | . | . | . | . | . |
| Carex pulicaris | 38– 1 | 1 | 1 | 1 | . | 1 | . | . | . | . | . | . | . |
| Viola canina | . | . | 2 | . | 2 | . | . | . | . | . | . | . | . |
| Festuca ovina | 88–35 | 25 | 25 | 30 | 15 | . | 10 | 10 | . | . | . | . | . |
| Anthoxanthum odoratum | 25– 1 | . | 1 | . | 5 | . | 3 | . | . | . | . | . | . |
| Carex flacca | . | 10 | 2 | 3 | . | 3 | 2 | . | . | . | . | . | . |
| Centaurea jacea | 75– 8 | 15 | . | . | . | . | 10 | . | . | . | . | . | . |
| Filipendula vulgaris | 38– 3 | 1 | 1 | . | 1 | 1 | . | 1 | . | . | . | . | . |
| Lathyrus pratensis | 50– 2 | 1 | . | 1 | . | 1 | 1 | 5 | . | . | . | . | . |
| Scorzonera humilis | 25– 2 | . | 2 | . | 1 | 1 | 1 | 10 | . | . | . | . | . |
| Succisa pratensis | 50– 8 | 15 | 20 | 1 | 10 | 1 | 5 | 5 | . | . | . | . | . |
| Dicranum scoparium | 62–15 | . | 70 | 3 | 50 | . | . | . | . | . | . | 1 | . |
| Carex nigra | 38– 1 | 5 | 2 | 10 | 1 | . | 1 | 10 | 2 | . | . | . | . |
| Carex panicea | 100– 4 | 20 | 3 | 10 | 3 | 10 | 2 | 2 | 1 | . | 1 | . | . |
| Salix repens | 38–37 | 20 | . | . | . | . | . | . | . | . | 1 | . | . |
| Carex | 12– 1 | . | 1 | . | . | . | 1 | . | . | 1 | . | . | . |
| Rumex acetosa | . | . | 2 | . | 1 | . | . | 5 | . | . | . | . | . |
| Climacium dendroides | 25– 1 | . | 1 | . | . | 15 | . | 1 | . | . | . | . | . |
| Valeriana dioica | 75– 2 | . | . | . | . | 5 | 20 | 5 | . | . | . | . | . |
| Aulacomnium palustre | 75–19 | . | . | . | 40 | . | . | . | . | 1 | . | . | . |
| Sieglingia decumbens | 62– 5 | 5 | 3 | 3 | . | . | 1 | 2 | . | . | 3 | . | . |
| Geum rivale | 50– 8 | 10 | 10 | 1 | 1 | 50 | 30 | 15 | 5 | 1 | . | . | . |
| Molinia caerulea | 100–15 | 10 | 15 | 25 | . | . | 30 | 3 | 15 | 1 | 5 | . | . |
| Potentilla erecta | 100–14 | 5 | 3 | 3 | 3 | 2 | 2 | 5 | 15 | 2 | 3 | 3 | . |
| Serratula tinctoria | 100– 7 | 5 | . | 10 | 3 | 5 | 25 | 5 | 3 | 2 | 2 | . | . |
| Thuidium tamariscinum | 62–15 | 70 | . | 15 | . | 5 | 1 | 1 | . | 60 | 5 | 3 | . |
| Lophocolea heterophylla | . | . | . | . | . | 1 | . | . | 1 | . | . | . | . |
| Galium boreale | 88– 2 | 2 | 2 | 2 | 1 | 20 | 5 | 10 | 2 | 5 | 5 | . | . |
| Equisetum arvense | 12– 1 | 2 | 1 | . | 1 | 2 | . | 1 | . | . | 15 | . | . |
| Carex disticha | 38– 2 | 1 | . | 5 | 3 | 1 | 1 | 2 | 3 | 1 | . | 2 | 100–24 |
| Vicia cracca | 38– 1 | . | 1 | . | . | . | . | . | . | 2 | 1 | . | . |
| Selinum carvifolia | . | 5 | 5 | 3 | 1 | 5 | 10 | 10 | 10 | 1 | 25 | 1 | . |
| Carex hirta | . | . | . | . | . | 2 | . | 2 | 2 | . | 1 | . | . |
| Festuca rubra | 25– 2 | . | . | . | 1 | 5 | 3 | 1 | . | 3 | 2 | . | . |
| Cirsium palustre | 12– 1 | . | 5 | . | . | 2 | . | 3 | . | 2 | . | 3 | 20– 4 |
| Filipendula ulmaria | 12– 1 | 2 | 2 | 5 | 1 | 50 | 70 | 80 | 80 | 99 | 99 | 99 | 90–83 |
| Brachythecium curtum | . | . | 1 | . | . | . | . | . | 3 | 2 | . | 10 | . |
| Calliergonella cuspidata | 25– 1 | . | . | . | . | . | . | . | 1 | 1 | . | . | . |
| Mnium affine | 25– 1 | . | . | . | 1 | . | . | 1 | 1 | . | . | 1 | . |
| Fissidens adianthoides | 50– 1 | . | . | . | . | 1 | . | 1 | 1 | . | . | 1 | . |
| Ranunculus auricomus | . | . | . | . | . | . | 2 | . | . | . | . | 1 | . |
| Crataegus | . | . | . | . | . | . | . | . | . | 60 | . | . | . |
| Salix aurita | . | . | . | . | . | . | . | . | . | 10 | . | 30 | . |
| Galium uliginosum | . | . | . | . | . | . | . | 1 | . | 1 | . | 1 | . |

128

## Discussion

### The formation and significance of tussocks

Many factors play a part in tussock formation. The texture and structure of the soil is important. As a rule sand does not provide the necessary aggregation; loam is more favourable. More stable tussocks are formed if a peaty substrate has developed in the uppermost 5-15 cm. Water erosion in slopes may leave tussocks standing. Some grasses (e.g. *Molinia caerulea*) also form tussocks and hold them together.

The development of tussocks in marshes on the slopes of Upper Teesdale was considered to be residual pieces of continuous turf, split by cattle-trampling and water erosion (Pigott 1956). The tussocks at Örup were perhaps developed in a similar way from continuous turf in the hay-meadows of the 18th century. Anyway, these tussocks do not seem to be generally dependent on the growth form of certain species, as do the tussocks of *Schoenus* fens, which usually occur independently of grazing (Tyler 1980).

Evidently the grazing and trampling of cattle is essential to maintain the tussocky vegetation. In the ungrazed parts almost nothing can be seen of tussocks. It seems that the tall species dominating these parts have replaced the shorter ones which once formed the firm tissue of the tussocks.

Wherever they develop, tussocks are an important differentiating factor of the habitat. Light and temperature conditions vary greatly between tussock tops, sides (especially the northern side), and depressions. Tussocks, once formed, also govern the trampling pattern of the cattle; out of a total 117 'steps', observed on freely grazing cattle, 106 were placed in depressions, 5 in an intermediate position, and only 6 on tussocks. Thus tussocks are perpetuated. It also leads to a considerable compactness in the depressions.

From a very different ecosystem Stuart Chapin, van Cleve & Chapin (1979) give a good example of the great ecological differences between tussocks and depressions. Of course the details concerning tussocks of *Eriophorum vaginatum* in Alaska are not relevant here. However, the general conclusions drawn fit very well into the picture, important points being

(1) The tussocks become snow-free earlier and experience a longer growing season. - At Örup the depressions are usually filled with ice or snow in winter and with water during the thaw while the tussocks only receive a fluffy covering of snow, which easily melts in spring.

(2) Tussocks thaw faster and become warmer in summer.

(3) Tussock plants have a deeper organic horizon to ex-

ploit. - This is also the case at Örup. The organic component is seldom detectable deeper than ca. 10 cm below the depressions.

(4) Tussock plants have a much higher leaf biomass than the plants in the depressions. - This is not always true at Örup, as the cattle prefer to graze on the tussocks. However, the productivity, judged in the field, seems to be higher on the tussocks.

There are also more or less marked pH-differences, the tussock tops being more acid. The relation between pH and bryophyte zonations on tussocks in *Schoenus* vegetation in Sweden has been thoroughly studied by Tyler 1980). She emphasizes the rôle of the subsoil water, as even a short-time rise may increase the pH in those parts of a tussock affected by the water.

At Örup every quadrat on the grazed side was categorized as tussock, depression or intermediate in relation to the immediate surroundings. Both the % and the HULT ordination clearly separate these categories from each other. A tendency to overlap can be observed among some of the slightly grazed quadrats where *Filipendula ulmaria* is expanding, gradually reducing the differences between tussocks and depressions. The results of the levelling show a much poorer connection with the vegetation (Fig. 9). A typical tussock may be at a lower level than a depression some dozen metres away.

### The succession

When grazing ceases, a vegetation of tall herbs and sedges starts to develop. The decomposition of litter does not keep pace with the production; the litter layer is relatively thick and evenly distributed in the ungrazed plots B and D. Only a few common species dominate. *Carex disticha* reigns in the somewhat wetter plot B, while *Filipendula ulmaria*, *Carex disticha* and *Urtica dioica* codominate in D. *Filipendula ulmaria* and *Carex disticha* are normal constituents of the grazed phase (in low abundance) but *Urtica dioica* is a clear sign of degeneration. All the dry habitat species and most of the bryophytes disappear.

The quadrats of the slightly grazed plot C have an intermediate position in the succession. The quadrats on the tussocks extend along a 'succession gradient' in the ordination, some being very similar to the heavily grazed tussocks of plot A (bottom left in Figs 1 & 2), others resembling the dense *Filipendula ulmaria*-dominated vegetation of plot D (to the right). This depends on a true patchiness in plot C, where some parts are still quite as heavily grazed as is plot A, while the vegetation in other parts changes. Cattle, well-known for their conservative habits, keep to the same tracks and open patches, where the grass, between the *Filipendula* stands, stays tasty. The first succession stages, where the cover of *Filipendula ulmaria* is moderate (20–40 %), seem especially unstable. This cover might be sufficient to discourage the grazing cattle and so the succession soon proceeds towards a more marked dominance of *Filipendula ulmaria*.

The most striking change in the succession on tussocks is the increase of *Filipendula ulmaria* which is accompanied by an increasing height in the vegetation and an accumulation of litter (Table 3). The number of species decreases. *Festuca ovina* and *Succisa pratensis*, important species of the tussocks, disappear as well as many others: *Briza media*, *Luzula multiflora*, *Carex pulicaris*, *Anthoxanthum odoratum*, *Carex flacca*, *Filipendula vulgaris*, *Scorzonera humilis* and even the rather stout *Centaurea jacea* and the often successfully climbing *Lathyrus pratensis*. *Carex nigra*, *C. panicea* and *Molinia caerulea* persist rather longer, however.

Few species other than *Filipendula ulmaria* gain from the changes. *Galium uliginosum*, usually occurring in the depressions, seems to increase, once the tussocks become shaded. Of great importance for the future succession is the establishment of small bushes such as *Salix aurita* and *Crataegus spp.*

It should be stressed that the situation in plot D, although in a way representing an extreme result of the succession proceeding in plot C, has not developed under the same conditions. The effects of a gradual reduction in grazing intensity are not the same as those of a sudden break. No woody plants have managed to enter plot D. A certain degree of grazing certainly favours such an establishment. The trampling breaks up the litter layer so that diaspores reach the soil surface. On the other hand the cattle often demolish the small bushes, especially those without thorns or spines.

## Summary

Two sites at Örup, SE Skåne, Sweden, have been investigated, viz. a grazed, unimproved, tussocky pasture on a calcareous moraine clay; and an originally similar, adjacent area that was abandoned about 1960. On the grazed site the vegetation is extremely rich in species. This vegetation type was formerly widespread, but nowadays it is rare and therefore it is important to try to conserve examples of it. In total, 80 0.1 m² quadrates distributed at random within 4 homogeneous plots were investigated. The species composition and the cover, the height of the vegetation, the position on a tussock or in a depression and the amount of litter were recorded. The cover data were ordinated (PCA). The differentiation between tussocks and depressions and the effects of ceasing grazing were clearly separated. The connections between the vegetation and the position of the quadrat, grazing intensity etc. were investigated. The vegetation of the ungrazed parts had become dominated by *Filipendula ulmaria* and *Carex disticha*; on an average there were 5.5 species/ 0.1 m². The grazed quadrats contained three times as many species and showed much higher spatial variation, important species being *Carex panicea*, *C. flacca*, *Molinia caerulea*, *Festuca ovina*, *Potentilla erecta*, *Centaurea jacea* and *Serratula tinctoria*.

## References

Arnell, S. 1956. Illustrated moss flora of Fennoscandia. I. Hepaticae. Lund, 308 pp.

Du Rietz, G.E. 1921. Zur methodologischen Grundlage der modernen Pflanzensoziologie. Uppsala, Wien, 267 pp.

Jensén, S. 1978. Influences of transformation of cover values on classification and ordination of vegetation. Vegetatio 37: 19–31.

Klika, J. 1934. Die Pflanzengesellschaften auf Travertinen bei Stankovany in der Slowakei. Bull. int. Acad. tchéque Sci. 35: 40–44. Prag.

Lid, J. 1974. Norsk og svensk flora. Oslo, 808 pp.

Malmer, N. 1962. Studies on mire vegetation in the Archaean area of Southwestern Götaland (South Sweden). 1. Vegetation and habitat conditions on the Åkhult mire. Op. Bot. Soc. Bot. Lund. 7: 1, 322 pp.

Nyholm, E. 1954–69. Illustrated moss flora of Fennoscandia. II. Musci. 1–6. Lund, 799 pp.

Persson, S. 1978. Datorprogram för bearbetning av ordinationsdata. 2. Ordinationsprogram – dokumentation och handhavande. Meddn Avd. Ekol. Bot., Lunds Univ. 34. 64 pp. Lund.

Pigott, C.D. 1956. The vegetation of Upper Teesdale in the North Pennines. J. Ecol. 44: 545–586.

Regnéll, G. 1976. Den sydsvenska kalkfuktängen i litteraturen. Meddn Avd. Ekol. Bot., Lunds Univ. 4: 3. 40 pp. Lund.

Regnéll, G. 1979. Vegetationsförändringar vid upphörande bete i en skånsk kalkfuktäng. Utvärdering och tolkningsmöjligheter med datateknik. Svensk Bot. Tidskr. 73: 139–159. Stockholm.

Roskam, E.E. 1971. Ordina: Multidimensional ordination of observation vectors. Program-Bulletin no 16. Psychological Laboratory, Catholic University Nijmegen, The Netherlands. 8 pp.

Stuart Chapin III, F., K. van Cleve & M.C. Chapin. 1979. Soil temperature and nutrient cycling in the tussock growth form of Eriophorum vaginatum. J. Ecol. 67: 169–189.

Tyler, C. 1980. Soil acidity and distribution of species on tussocks and interspaces in Schoenus vegetation of South and Southeast Sweden. Vegetatio (In press).

Westhoff, V. & A.J. den Held. 1969. Plantengemeenschappen in Nederland. Zutphen, 324 pp.

Accepted 14 December 1979

# SUCCESSION: A POPULATION PROCESS

Robert K. PEET[1] & Norman L. CHRISTENSEN[2]

[1] Department of Botany, University of North Carolina, Chapel Hill, North Carolina, 27514 USA*
[2] Department of Botany, Duke University, Durham, North Carolina 27706 USA

Keywords:

Forests, Mortality, North Carolina, Plant demography, Succession, Thinning, Vegetation

## Introduction

For over 80 years succession theory has played a central role in plant ecology, providing both a predictive tool and organizational scheme. Drawing on this long history of research and observation, Margalef (1968), Odum (1969), Whittaker (1975) and others have identified general trends in community development. Their synthetic treatments have helped focus the efforts of subsequent workers examining the empirical and experimental basis of succession theory (e.g. Connell & Slatyer 1977, Drury & Nisbet 1973, Egler 1975, Horn 1974, Pickett 1976, van Hulst 1978). This more recent work suggests that the classical succession paradigm is seriously flawed and that many long held concepts need to be reexamined.

The view of succession as a community or species replacement sequence driven by autogenic environmental modification has been rejected (Connell & Slatyer 1977, Drury & Nisbet 1973, Egler 1954, 1976, Niering & Egler 1955). In contrast to Odum's (1969) generalizations, successional changes in biomass and primary production can no longer be assumed to be consistently upward. Bormann & Likens (1979), Horn (1974), Loucks (1970), Major (1974) and Peet (1978, 1980a) all suggest that a more usual pattern is to find maximum productivity and biomass preceding the climax stage of forest development. Also in opposition to Odum's generalizations, Vitousek & Reiners (1975, Vitousek 1977) and Bormann & Likens (1979) suggest that the tightest nutrient cycles should be expected in the central portion of successional sequences. A similar reevaluation has proven necessary for patterns

* This research was supported by National Science Foundation grants DEB-7708743 and DEB-7804043 to R.K.P. and DEB-7707532 and DEB-7804041 to N.L.C.

in species diversity (Auclair & Goff 1971, Loucks 1970, Peet 1978, Whittaker 1972, 1977).

Simultaneous with the realization that succession is far more complex than classical theory suggests, there appeared in the literature a series of alternative approaches and formulations. These were not proposed as mutually exclusive, competing hypotheses, and should not be taken as such. Rather, they may all apply in varying degrees to any one successional sequence. What is of interest is the extent to which they are successful, either separately or in combination, in increasing our understanding of the succession process. The most important and frequently cited of these alternative approaches can be summarized in three groups.

1. *Succession as a gradient in time:* Drury & Nisbet (1971, 1973) argued that a successsional sequence is simply a type of stress gradient to which plants are adapted. Their viewpoint was reductionistic, almost Gleasonian, and they suggested that ' . . . a complete theory of vegetational succession should be sought at the organismic, physiological or cellular level and not in emergent properties of populations or communities.' Pickett (1976) also viewed the successional sequence as a form of gradient, along which species are competitively displaced. From this viewpoint the physiological and life-history characteristics of the component species are of central importance. 'Succession,' Picket wrote, 'can be understood solely in terms of the interaction of evolutionary strategies without reference to a deterministic progress toward climax.' It is precisely this evolutionary strategy, life-history approach which Noble & Slatyer (1977, 1980) have been employing in their efforts to identify a minimal number of 'vital attributes' of plant species necessary to predict successional changes.

2. *Differential longevity:* Egler & Niering (1955, Egler 1954, 1976) have argued that much of secondary succession is simply a consequence of differential longevity, that most of the eventual dominants enter a community in the earliest developmental stages when competitive pressures are low. Drury & Nisbet (1973) also considered initial conditions and species composition critical determinants of future community development. These arguments imply that failure of a species to become established early prevents, or at least greatly reduces its chances for subsequent dominance. Connell & Slatyer (1977) referred to this as the inhibition model in contrast to the classical facilitation model of Clements and others where in one species or set of species prepares the way for the next. If history is as important for vegetation development as Drury & Nisbet, Egler, and others suggest, convergence to climax may be too slow to be of practical significance, and perhaps does not occur at all.

3. *Succession as a stochastic process:* Horn (1975, 1976) has presented a model based on tree-by-tree replacement which has greatly popularized Markovian models. Central to every discrete Markov model is a matrix of transition probabilities. In Horn's model the matrix consists of the probabilities of a tree of any given species being replaced by a tree of the same species or any other species. Such a matrix can be used to calculate future forest composition and in this way provides a useful neutral model against which to compare reality. A necessary statistical consequence of this formulation is that convergence to a steady-state composition in inevitable, regardless of initial conditions. The interesting biology is, thus, not in the existence or even composition of the climax, but in the numbers comprising the transition matrix, especially their variation with site and stand development. Stochastic succession models need not be restricted to Markovian tree-by-tree replacement processes, but may be used at various hierarchical levels from trees of varying size-classes (e.g. Usher 1966, Moser 1972) to regional forest-types (e.g. Shugart et al. 1973), or may even be based on complex tree population simulations (e.g. Botkin et al. 1972).

Despite obvious differences, a common theme characterizes the preceding approaches; all are reductionist, emphasizing life-histories and competitive relations of component species rather than the emergent properties of communities. To date few data have been gathered appropriate for evaluating their relative utility or generality.

The intent of this contribution is not to further review previous studies, but to preview a research program we have undertaken to evaluate in-depth, for one area, the utility of a reductionist, population-based approach to studying succession. Our approach is to view succession as a population process – a consequence of variation in rates of reproduction, establishment, growth and mortality. Species life-history and physiological characteristics determine, to a large extent, potential population responses to the changing competitive environment. Population processes in turn appear to strongly influence such community-level properties as species composition, rate of compositional change, diversity, productivity and biomass. Thus, we are taking a two-step or three-tier reductionist approach to the study of succession.

### The study area

Our study is being conducted primarily in the Duke Forest, an experimental forest owned by Duke University and located on the North Carolina piedmont. The legacy of prior research on the study area significantly influenced our study area selection. The papers of Billings (1938), Korstian & Coile (1938), Kozlowski (1949), Oosting (1942) and many others all have been based on studies conducted within the Duke Forest. Information on the physical environment of the area can be found in Billings (1938) and Oosting (1942).

Virtually all North Carolina piedmont vegetation is either the direct result of or has been greatly modified by human activities during the last 200 years. Most of the piedmont was at one time under cultivation, but changing economic conditions led to abandonment of much of this land and its reversion to forest. Typically, initial post-abandonment forest is dominated by pines; primarily *Pinus taeda* or *P. echinata*, or farther west, *P. virginiana*. All three species of *Pinus* occur within the study area and broadly overlap in their ecological requirements, but with *P. taeda* dominant on more mesic, nutrient rich sites and *P. virginiana* dominating a few of the edaphically least favorable sites. A number of light-demanding hardwood species often invade simultaneously with or slightly after the pines. The more rapidly growing of these species, especially *Liquidambar styraciflua* and *Liriodendron tulipifera*, occasionally codominate with pine, but are not commonly present in large numbers in the canopy.

The *Pinus* species in our region are light requiring and cannot regenerate successfully without large canopy gaps such as created by selective cutting or severe wind.

In contrast, a number of hardwood species are capable of invading beneath the pines. As a result, 80 to 100 years after land abandonment when the pine canopy is breaking up through natural mortality, numerous young hardwood saplings are available to fill the gaps. The composition of the new hardwood forest will probably change gradually for another 200 to 500 years as early invaders such as *Liriodendron*, *Liquidambar*, and *Acer rubrum* are replaced by more slowly invading species, especially species of *Quercus* and *Carya*. A more detailed description of the potential steady-state vegetation and its variation with site conditions can be found in Peet & Christensen (1979, Peet 1980b.)

## Methods

The study of plant population dynamics is best accomplished through long-term observations of marked individuals under carefully designed experimental conditions. As Harper (1977) has emphasized, the large size and great longevity of trees has largely precluded such critical demographic studies. Fortunately, we are heir to a set of per-

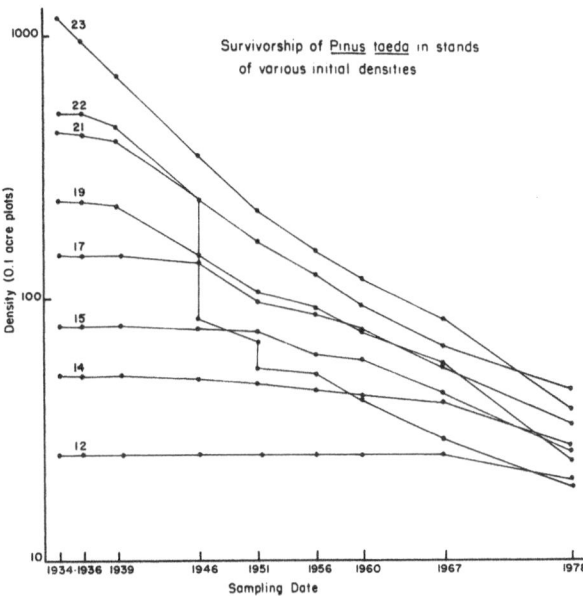

Fig. 1. Survivorship of marked *Pinus taeda* ( > 1.25 cm dbh) in stands of differing initial densities over a 44 year period. Initial densities of trees in the 0.1 acre (.0405 ha) plots were 25 (plot 12), 51 (14), 79 (15), 149 (17), 234 (19), 431 (21), 511 (22), and 1172 (23). Sample plot 22 was twice thinned as represented by the vertical portions of its survivorship curve. Despite extreme variation in initial densities, differential mortality has led to convergence in density during the 44 year period.

manent sample plots established within the Duke Forest during the early 1930's. In each plot the diameter and height of each tree was measured at roughly 5 year intervals until 1963, and then again 15 years later. Plots were typically established in sets with one or two designated as controls and the others subjected to various experimental regimes.

To illustrate the potential of a population-based approach for studying succession we have selected as examples three sets of permanent sample plots. The first set (12 – 23) started with twelve 8-year-old *Pinus taeda* plots of 0.1 acre (.04 ha) on a single hillside, but ranging in natural initial density from 25 to 1143. Seven of these were left as controls (Fig. 1). The second set (24–26) consists of three matched 19-year-old *Pinus taeda* plots which received different treatments. One plot (24) was 'thinned from above,' removing only the largest trees. The second plot (25) was 'thinned from below,' removing overtopped and intermediate trees. Both plots were thinned to a basal area of roughly 23 m²/ha (100 ft²/acre). The third plot (26) was retained as a control (initial basal area of .29.8 m²/ha). The third set (10,36,37) consists of three 0.25 acre (0.1 ha) plots dominated by mixed hardwoods, primarily species of *Quercus* and *Carya*. Plots 36 and 37 were similar except that 37 was significantly thinned by a hurricane in 1954 which reduced the basal area from 25 to 15 m²/ha, thus opening the canopy. Plot 10 differs from 36 in that it appears to represent an earlier successional stage, probably resulting from selective cutting in the late 19th century. Evidence for this is seen in several aspects of the 1934 data: basal area was lower (18.8 vs. 22.0 m²/ha), average diameter was lower (10.3 vs. 13.8 cm), average height was lower (9.8 vs. 11.5m) and density was higher (3152 vs. 1432 stems/ha).

Unfortunately the foresters who established these permanent plots were little interested in tree seedlings and initial tree establishment; only individuals over 1.25 cm dbh were measured. Therefore we have initiated a series of permanent plots for monitoring seedling establishment and survival. However, several more years of observation will be required before these can contribute substantially to our understanding of tree demography and successional processes.

## Results

### Mortality and thinning

Tree mortality is a process central to any consideration of

133

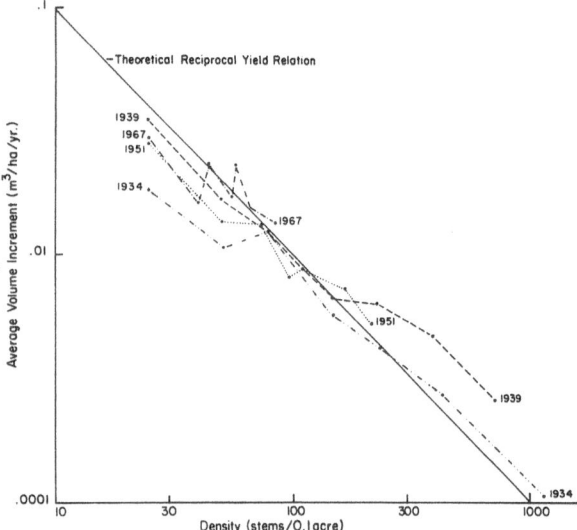

Fig. 2. Logarithmic graphs of average annual volume increment (m³ ha⁻¹ yr⁻¹ versus stand density for the seven control stands illustrated in Figure 1. A reciprocal annual yield relationship would conform to a − 1 slope. Despite considerable variation in tree age, density and size among stands and among years, strong conformity to the reciprocal yield relation can be seen.

forest compositional change. Using the examples we have selected, aspects of mortality in both the *Pinus taeda* stage and the mixed hardwood stage can be illustrated.

Tree density strongly influences mortality as can be seen in Fig. 1. In each of the 7 matched stands, after a critical degree of crowding was reached, mortality showed a nearly perfect exponential decrease. (Lack of data for stands less than 8-year-old could partially explain failure of the data to fit either the power function suggested by Hett (1971) or the sigmoidal relation suggested by Goff & West (1975). Despite the broad dispersion of initial densities (25–1143/0.1 acre plot = 618–28244/ha) the plots converged within 45 years to a rather narrow range of densities (20–45/plot). In contrast, the experimentally thinned plot (22) never recovered from its initially high density; the initial high mortality rate continued after competitive release. Most likely some aspect of tree geometry was fixed by initial growing conditions. Our studies of the growth characteristics of *Pinus taeda* suggest that in dense populations a high ratio of height to diameter growth occurs with the resultant tall, thin trees being highly susceptible to wind and ice damage.

Given stands of markedly differing densities and developmental stages, such as in plots 12–23, it is interesting to consider the degree of their convergence in production.

As densities were sufficiently high to result in density dependent mortality, a reciprocal yield relationship (Harper 1977, Kira et al. 1953) was expected. Specifically, the reciprocal yield relation suggests that total productivity should be independent of density. Alternatively stated, a double logarithmic graph of average tree productivity versus density should yield a plot with a slope of − 1. This relation has not to our knowledge been tested elsewhere for trees. Such a test must differ from work on herbaceous species; herbs can simply be harvested and weighed to provide a production estimate but this is not possible for trees. Instead, we have estimated net production using 'estimated volume increment' (Whittaker & Woodwell 1968, Whittaker & Marks 1975 – Assume a paraboloid of rotation based on tree height and diameter, and subtract a second such paraboloid calculated using height and diameter data from the previous year). The results, shown in Fig. 2, strongly support the applicability of the reciprocal yield relation to trees. Especially note-

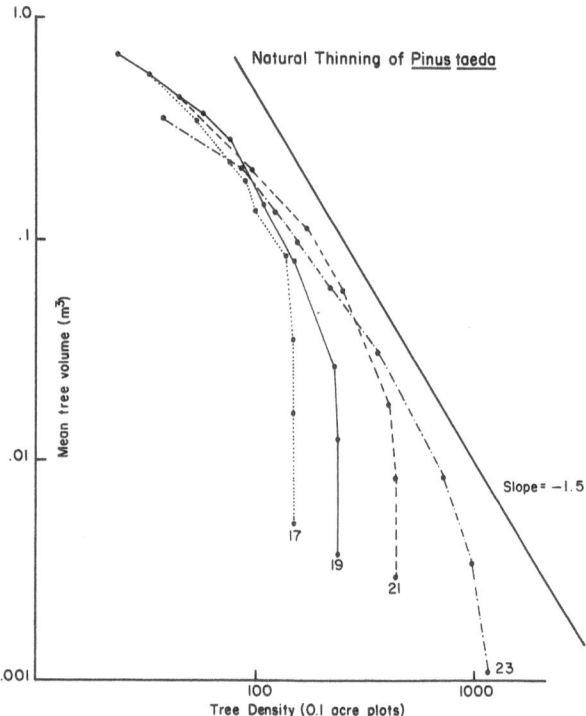

Fig. 3. Logarithmic plots of changing tree volume versus tree density for four sample plots of differing initial densities (149, 234, 431 and 1172 per 0.1 acre, .0405 ha) illustrate that Yoda's –3/2 law of thinning does not strictly apply to even-aged *Pinus taeda* stands; the shift toward a –3/2 relation is followed by a shift toward –1 as trees mature and reach limiting dimensions.

134

worthy is the independence of production and age of the stand.

The link between site conditions, density, and mortality can be further examined by constructing double logarithmic graphs of single stands through time showing density versus average tree biomass (Fig. 3). Convergence toward a line with a slope of $-3/2$ would imply agreement with Yoda et al.'s (1963) thinning law. This relation has only rarely been examined for time-series tree data (e.g. Drew & Flewelling 1977), and then with ambiguous results.

Our plots do not exhibit strong conformance with the Yoda relation. Rather, the plots at first shift toward slopes of $-3/2$, but continue past eventually approaching slopes of $-1$. This could be a result of trees reaching their vertical growth potential well before diameter growth rates start to decrease.

Mortality data from uneven-aged, mixed hardwood stands are more ambiguous, primarily due to the unknown, ever-changing age structure of the populations, and the size dependence of mortality. Unless a stand can be shown to be at steady-state, which is rarely if ever the case, a depletion curve cannot strictly be interpreted as a survivorship curve. This limitation is important but does not prevent useful comparisons from being made.

Depletion curves for hardwoods alive in 1934 in plots 10 and 36 are shown in Fig. 4. As was the case for *Pinus taeda*, the curves conform well with a negative exponential model. Comparing the two stands, the difference in the overall or total depletion rate is conspicuous. A

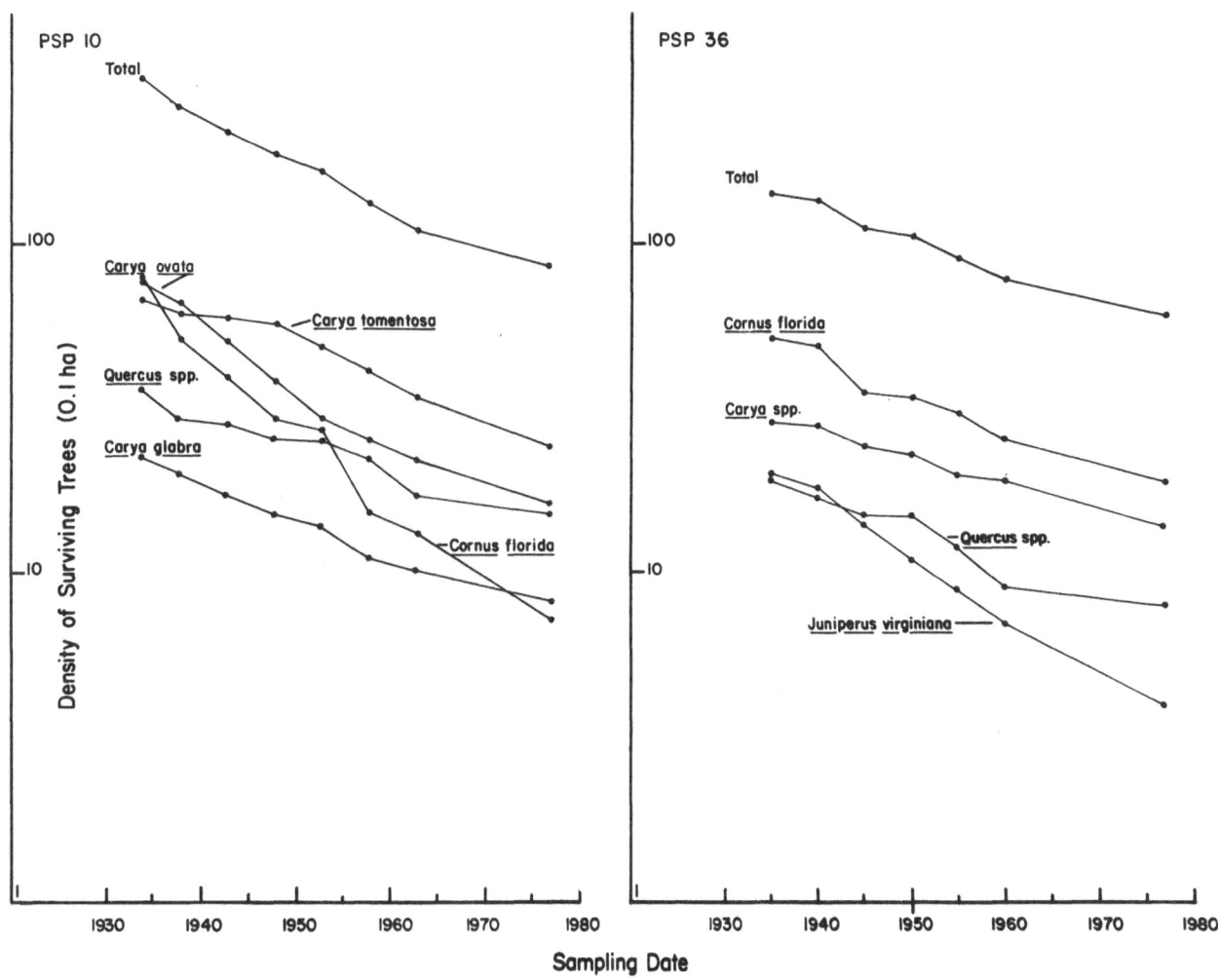

Fig. 4. Depletion curves illustrating differential mortality in trees from two upland, uneven-aged, mixed-species stands of differing sucessional status. The less mature plot 10 can be recognized by the greater rates of thinning and tree mortality (see text).

135

graph of the logarithm of stem density (stems $\geq 1.25$ cm/ha) versus time (in centuries) yields a slope of $-1.42$ for plot 10 compared to only $-1.02$ for plot 36. The greater depletion rate in plot 10 may be symptomatic of its presumed earlier successional age where thinning processes dominate to a greater extent. In addition, species differences can be seen both within and between stands. Within plot 10 the three *Carya* species can be seen to be responding quite differently. *Cornus florida* appears as an average species in plot 36, but in plot 10 it has the highest depletion rate of any of the common species.

*Establishment*

Complementary to tree mortality is establishment of new individuals. As future forest composition and structure are directly dependent on new tree establishment, an experimental examination of this process is critical for understanding succession.

A set of three permanent sample plots in what was initially a 19-year-old *Pinus taeda* forest provides an example of this experimental approach. In one plot the dominant trees were selectively cut (thinned from above),

in a second the suppressed understory trees were removed (thinned from below), and a third plot was left as a control. Table 1 shows stand composition after the initial treatment, and again 43 years later. The density of hardwoods ($\geq 1.25$ cm dbh) in the control plot increased by a factor of 2.8 despite mortality of almost half of the stems alive at the start of the period. In the plot thinned from below most of the hardwoods were removed in the initial thinning producing an artifically low initial density. Nonetheless, hardwood density after 43 years was roughly equivalent to that on the control plot, an increase of 1440%. The greater age and size of the control hardwoods predictably led to higher basal area on that plot. In contrast to these two plots, the increase in hardwood density in the plot thinned from above was small, the final hardwood density being less than half that on either of the others. Greatest production, as indicated by basal area increment, occurred in the plot thinned from below, while the plot thinned from above had the lowest increment.

A major part of the increase in hardwood density in all the stands occurred in predominantly understory species such as *Cornus florida*, *Oxydendrum arboreum*, and *Acer rubrum*. *Nyssa sylvatica* and *Liquidambar* are less

Table 1. Effect of thinning on hardwood invasion of *Pinus taeda* forests.

| Species          Year | Thinned from above | | | Thinned from below | | | Control | | |
|---|---|---|---|---|---|---|---|---|---|
|  | 1934 | 1979[2] | 1979+[2] | 1934 | 1979 | 1979+ | 1934 | 1977 | 1977+ |
| Acer rubrum | 20[1] | 10 | 306 | 0 | 0 | 928 | 0 | 0 | 503 |
| Carya spp. | 267 | 100 | 100 | 20 | 10 | 50 | 60 | 30 | 30 |
| Cornus florida | 10 | 10 | 60 | 20 | 0 | 424 | 50 | 30 | 642 |
| Liquidambar | 168 | 90 | 405 | 60 | 30 | 494 | 415 | 227 | 721 |
| Nyssa sylvatica | 80 | 60 | 109 | 80 | 10 | 119 | 178 | 119 | 217 |
| Oxydendrum | 50 | 10 | 90 | 0 | 0 | 50 | 148 | 108 | 395 |
| Pinus spp. | 2510 | 435 | 464 | 1047 | 375 | 375 | 2311 | 355 | 355 |
| Quercus spp. | 0 | 0 | 30 | 0 | 0 | 100 | 0 | 0 | 40 |
| Other | 100 | 20 | 213 | 10 | 0 | 572 | 110 | 0 | 139 |
| Total Density | 3205 | 735 | 1773 | 1237 | 425 | 3112 | 3272 | 869 | 3042 |
| Hardwood Density | 695 | 300 | 1309 | 190 | 50 | 2737 | 961 | 514 | 2687 |
| Total Basal Area | 23.4[3] | 29.5 | 32.0 | 22.8 | 39.2 | 44.5 | 30.8 | 35.8 | 40.9 |
| Hardwood B.A. | 0.9 | 1.9 | 4.3 | 0.2 | 0.7 | 5.9 | 1.2 | 4.4 | 9.6 |

[1] Densities in stems $\geq 1.25$ cm dbh per ha.
[2] 1979 values contain only stems $\geq 1.25$ cm dbh in 1934 surviving to 1979.
1979 + values include all stems $\geq 1.25$ cm dbh in 1979.
[3] m$^2$/ha.

136

Table 2. Effect of thinning on tree establishment in upland hardwood forests.

A. Control plot

| Diameter Class[1] | 1934 | | | 1950 | | | 1977 | | |
|---|---|---|---|---|---|---|---|---|---|
| | 1 | 2 | 3 | 1 | 2 | 3 | 1 | 2 | 3 |
| Oxydendrum a. | 1[2] | 11 | 5 | 3 | 10 | 6 | 5 | 10 | 8 |
| Quercus velutina | 0 | 1 | 2 | 0 | 0 | 3 | 0 | 0 | 2 |
| Carya spp. | 5 | 6 | 18 | 1 | 3 | 16 | 0 | 1 | 14 |
| Juniperus v. | 9 | 8 | 0 | 2 | 9 | 1 | 1 | 1 | 3 |
| Cornus florida | 7 | 43 | 2 | 12 | 27 | 6 | 46 | 22 | 3 |
| Quercus coccinea | 0 | 2 | 6 | 0 | 0 | 6 | 1 | 0 | 1 |
| Quercus alba | 1 | 1 | 5 | 0 | 1 | 5 | 0 | 2 | 4 |
| Acer rubrum | 0 | 1 | 1 | 1 | 1 | 1 | 4 | 0 | 2 |
| Liriodendron t. | 0 | 0 | 2 | 0 | 0 | 2 | 0 | 0 | 2 |
| Nyssa sylvatica | 0 | 3 | 0 | 0 | 0 | 2 | 0 | 0 | 2 |
| Quercus stellata | | | | | | | 1 | 0 | 0 |
| Quercus rubra | | | | | | | 1 | 0 | 0 |
| Carpinus | | | | | | | 3 | 0 | 0 |
| TOTAL | 23 | 76 | 41 | 21 | 54 | 46 | 65 | 36 | 40 |

B. Thinned plot (1954 hurricane)

| Diameter Class[1] | 1934 | | | 1950 | | | 1977 | | |
|---|---|---|---|---|---|---|---|---|---|
| | 1 | 2 | 3 | 1 | 2 | 3 | 1 | 2 | 3 |
| Carya spp. | 2[2] | 13 | 2 | 0 | 7 | 3 | 75 | 2 | 0 |
| Cornus florida | 3 | 25 | 1 | 0 | 26 | 2 | 29 | 24 | 1 |
| Quercus alba | 0 | 5 | 23 | 0 | 2 | 23 | 11 | 1 | 0 |
| Juniperus v. | 3 | 2 | 1 | 2 | 2 | 1 | 13 | 13 | 0 |
| Oxydendrum a. | 0 | 6 | 2 | 2 | 2 | 3 | 31 | 10 | 1 |
| Quercus rubra | 0 | 2 | 0 | 0 | 1 | 1 | 1 | 0 | 1 |
| Acer rubrum | 0 | 1 | 0 | 6 | 1 | 0 | 92 | 16 | 11 |
| Nyssa sylvatica | 0 | 1 | 0 | 0 | 1 | 0 | 4 | 1 | 0 |
| Ulmus alata | | | | | | | 1 | 0 | 0 |
| Pinus virginiana | | | | | | | 1 | 12 | 2 |
| Pinus taeda | | | | | | | 5 | 27 | 1 |
| Sassafras a. | | | | | | | 16 | 6 | 0 |
| Prunus serotina | | | | | | | 9 | 2 | 0 |
| Diospyros v. | | | | | | | 4 | 1 | 0 |
| Quercus velutina | | | | | | | 21 | 4 | 0 |
| Ostrya v. | | | | | | | 1 | 2 | 0 |
| Liriodendron t. | | | | | | | 20 | 12 | 0 |
| Chionanthus v. | | | | | | | 1 | 0 | 0 |
| TOTAL | 8 | 55 | 30 | 10 | 42 | 33 | 335 | 133 | 17 |

[1]Diameter classes of trees are (1) 2.5–5 cm, (2) 5–10 cm, and (3) > 10 cm.
[2]Densities are in stems per 0.1 ha.

shade tolerant species which usually become established early in succession, a fact which could account for their greater abundance in the control plot. Species of genera normally associated with the climax forests of the region, *Quercus* and *Carya* (see Oosting 1942), show very small levels of increase. A combination of the lack of nearby mature individuals and large seed size (i.e. poor dispersal) could largely explain their failure to reinvade.

The influence of thinning on mature hardwood forests (Table 2) is illustrated in a second example. A hurricane severely damaged one plot in 1954 decreasing its basal area from 25 m²/ha in 1950 to 15 m², whereas the control plot received only minor damage. Low levels of regeneration of all tree species except *Cornus florida* have been the norm for the control since 1934 whereas the experimental plot showed a dramatic increase in regeneration of most species following the hurricane. Of particular interest is the negligible level of *Carya*, *Quercus*, *Pinus* and *Liriodendron* regeneration in the control plot, and in the experimental plot prior to perturbation, an observation consistent with most reported studies of *Quercus-Carya* forests in eastern North America (e.g. Buell et al. 1966, Christensen 1977, Good 1965, Keever 1973, Peet & Loucks 1977). The obvious post-disturbance increase in regeneration of these species in the experimental plot suggests that a dynamic, temporally patchy type of forest regeneration characterizes these species in climax forests.

**Discussion and conclusions**

Establishment and mortality are complementary population processes which together directly determine changes in community composition and structure. Unfortunately, the difficulty of studying tree populations (see Harper 1977) has precluded all but a very few applications of population based approaches to forest succession research (e.g. Hartshorn 1975, Sarukhan 1978). Shortcuts such as inference from size structure have more often been employed (e.g. Peet & Loucks 1977). For example, pine replacement by hardwoods is anticipated from comparisons of saplings with mature trees in Table 1. Similarly, compositional differences between saplings and mature trees in the mixed *Quercus*, *Carya* stands of Table 2 are suggestive of successional trends (see Christensen 1977). However, only through direct observation of permanent plots can structural properties, such as size distributions, be firmly linked with successional processes, and only through experimental manipulation of such plots can causal mechanisms be identified.

In some stages of forest development mortality processes dominate. Changes in the *Pinus taeda* plots shown in Fig. 1 have been solely due to mortality. Similarly, mortality has been the dominant factor for over 45 years in the hardwood stands shown in Fig. 4. The reverse situation of the establishment process dominating forest change is represented by the hurricane damaged plot shown in Table 2b.

The experimental thinning treatments shown in Table 1 illustrate the shifting importance of mortality and establishment. While all three plots had marked *Pinus* mortality, the plot which was thinned from above was distinctive in its low rate of hardwood establishment. The experiment suggests that if a mature canopy is greatly disturbed, the understory saplings will be released to compete for canopy dominance. Intense competition should follow with little regeneration possible under the resulting dense, even-aged stand. This contrasts with the case of limited canopy damage represented by the plot thinned from below (or the hardwood plot thinned by the 1954 hurricane) which allowed increased regeneration, but not sufficient understory growth to quickly eliminate further establishment.

The preceding observations suggest a general pattern. Initially after severe disturbance the establishment process dominates. However, after a relatively short period, a dense stand of trees develops, precluding further establishment. From this point forest succession becomes a thinning process which lasts either until a subsequent disturbance, or until natural thinning opens the canopy and a balance is achieved between patches which are thinning and patches dominated by establishment. Parallel to the

compostional changes will be changes in such 'community properties' as biomass and diversity. Diversity should be minimal during the thinning stage when herbs, like small trees, are less common on the forest floor. Biomass will initially increase but restricted establishment will eventually cause it to plateau and then drop somewhat, allowing establishment to resume. Similar results have been suggested by Bormann & Likens (1979) and Peet (1978).

The alternative formulations of succession theory outlined in the introduction all fall into an establishment-mortality model. Connell & Slatyer's (1977) proposed facilitation and inhibition mechanisms apply strictly to the influence of canopy individuals on establishment. The differential longevity approach applies to the growth phase when mortality dominates. The idea of time as a gradient is easily applied with different species populations simply responding through variation in their mortality and establishment rates. Stochastic models can be based on mortality and establishment rates (see Botkin et al. 1972), though some of the simpler Markovian models proposed do not allow changes in rates over time, thus precluding their meaningful application to short-term successional situations.

**Summary**

Recent critical reviews suggest the need for a reductionistic approach to the study of secondary plant succession. We propose viewing succession as the result of the underlying plant population dynamics. This approach is being developed using nearly 50 years of permanent sample plot records.

After initial establishment *Pinus taeda* shows an exponential depletion with stands of various densities conforming to the reciprocal yield relationship. Uneven-aged hardwoods also show exponential depletion. Canopy disturbance can enhance the establishment process, though severe disturbance and the consequent abundant regeneration can lead again to dense, even-aged stands with low levels of establishment. These results suggest a general pattern of forest development wherein establishment is initially important, but is quickly replaced by mortality as the dominant process when the dense, even-sized stand starts to thin. Eventually, failing additional disturbance, natural mortality will again open the canopy allowing development of a balance between establishment and mortality.

## References

Auclair, A.N. & F.G. Goff. 1971. Diversity relations of upland forests in the Western Great Lakes area. Amer. Nat. 105: 499–528.

Billings, W.D. 1938. The structure and development of old field shortleaf pine stands and certain associated physical properties of the soil. Ecol. Monogr. 8: 437–499.

Bormann, F.H. & G.E. Likens. 1979. Pattern and process in a forested ecosystem. Springer-Verlag, N.Y., 253 pp.

Botkin, D.B., J.F. Janak & J.R. Wallis, 1972. Some ecological consequences of a computer model of forest growth. J. Ecol. 60: 849–872.

Buell, M.F., A.W. Langford, D.W. Davidson & L.F. Ohmann. 1966. The upland forest continuum in northern New Jersey. Ecology 47: 416–432.

Christensen, N.L. 1977. Changes in structure, pattern, and diversity associated with climax forest maturation in Piedmont, North Carolina. Amer. Midl. Nat. 97: 176–188.

Connell, J.H. & R.O. Slatyer. 1977. Mechanisms of succession in natural communities and their role in community stability and organization. Amer. Nat. 111: 1119–1144.

Drew, T.J. & J.W. Frewelling. 1977. Some recent Japanese theories of yield-density relationships and their application to Monterey Pine plantations. For. Sci. 23: 517–534.

Drury, W.H. & I.C.T. Nisbet. 1971. Inter-relations between developmental models in geomorphology, plant ecology and animal ecology. Gen. Sys. 16: 57–68.

Drury, W.H. & I.C.T. Nisbet. 1973. Succession. J. Arnold Arb. 54: 331–368.

Egler, F.E. 1954. Vegetation science concepts. 1. Initial floristic composition – a factor in old-field vegetation development. Vegetatio 4: 412–417.

Egler, F.E. 1976. Nature of vegetation. Its management and mismanagement. Conn. Cons. Assoc., Bridgewater, Conn., 527 pp.

Goff, F.G. & D. West. 1975. Canopy-understory interaction effects on forest population structure. For. Sci. 21: 98–108.

Good, N.F. 1968. A study of natural replacement in six stands in the highlands of New Jersey. Bull. Torrey Bot. Club 95: 240–253.

Harper, J.L. 1977. Population biology of plants. Academic Press, N.Y., 892 pp.

Hartshorn, G.S. 1975. A matrix model of tree population dynamics. In: F.B. Golley & E. Medina (eds.), Tropical ecological systems. p. 41–52, Springer-Verlag, N.Y.

Hett, J.M. 1971. A dynamic analysis of age in sugar maple seedlings. Ecology 52: 1071–1074.

Horn, H.S. 1974. The ecology of secondary succession. Ann. Rev. Ecol. Syst. 5: 25–37.

Horn, H.S. 1975. Markovian properties of forest succession. In: M.L. Cody & J.M. Diamond (eds.), Ecology and evolution of communities. p. 196–211, Belknap Press, Cambridge, Mass.

Horn, H.S. 1976. Succession. In: R.M. May (ed.), Theoretical ecology: principles and applications. p. 187–204. Blackwell, London.

Hulst, R. van. 1978. On the dynamics of vegetation: patterns of environmental and vegetational change. Vegetatio 38: 65–75.

Keever, C. 1973. Distribution of major forest species in southeastern Pennsylvania. Ecol. Monogr. 43: 303–327.

Kira, T., H. Ogawa & K. Shinozaki. 1953. Intraspecific competition among higher plants. 1. Competition-diversity-yield inter-relationships in regularly dispersed populations. J. Inst. Polytech. Osaka City Univ. D 4: 1–16.

Korstian, C.F. & T.S. Coile. 1938. Plant competition in forest stands. Duke Univ. For. Bull. 3. 125 pp.

Kozlowski, T.T. 1949. Light and water in relation to growth and competition of piedmont forest tree species. Ecol. Monogr. 19: 207–231.

Loucks, O.L. 1970. Evolution of diversity, efficiency, and community stability. Amer. Zool. 10: 17–25.

Major, J. 1974. Biomass accumulation in successions. In: R. Knapp (ed.), Vegetation dynamics, Handbk. Veg. Sci. 8: 195–203. Junk, The Hague.

Margalef, R. 1968. Perspectives in ecological theory. Univ. Chicago Press. 111 pp.

Moser, J.W. 1972. Dynamics of an uneven-aged forest stand. For. Sci. 18: 184–191.

Niering, W.A. & F.E. Egler. 1955. A shrub community of Viburnum lentago, stable for twenty-five years. Ecology 36: 356–360.

Noble, I.R. & R.O. Slatyer. 1977. Post-fire succession of plants in Mediterranean ecosystems. In: H.A. Mooney & C.E. Conrad (eds.), Symposium on the environmental consequences of fire and fuel management in Mediterranean ecosystems. p. 27–36. U.S.D.A. For. Serv. Gen. Tech. Rpt. WO-3.

Noble, I.R. & R.O. Slatyer. 1980. The use of vital attributes to predict successional changes in plant communities subject to recurrent disturbances. Vegetatio 43: ...–....

Odum, E.P. 1969. The strategy of ecosystem development. Science 164: 262–270.

Oosting, H.J. 1942. An ecological analysis of the plant communities of Piedmont, North Carolina. Amer. Midl. Nat. 28: 1–126.

Peet, R.K. 1978. Forest vegetation of the Colorado Front Range: Patterns of species diversity. Vegetatio 37: 65–78.

Peet, R.K. 1980a. Forest vegetation of the Northern Colorado Front Range, U.S.A.: II. Forest structure. Vegetatio (in press).

Peet, R.K. 1980b. Ordination as a tool for analyzing complex data sets. In: E. van der Maarel (ed.), Advances in vegetation science: Classification and Ordination, Vegetatio 42: ...–....

Peet, R.K. & N.L. Christensen. 1979. Hardwood forest vegetation of the North Carolina piedmont. Veröff. Geobot. Inst. ETH, Stiftung Rübel, 1979 (in press).

Peet, R.K. & O.L. Loucks. 1977. A gradient analysis of southern Wisconsin forests. Ecology 58: 485–499.

Pickett, S.T.A. 1976. Succession: an evolutionary interpretation. Amer. Natur. 110: 107–119.

Sarukhan, J. 1978. Studies on the demography of tropical trees. In: P.B. Tomlinson & M.H. Zimmermann (eds.), Tropical trees as living systems, p. 163–184. Cambridge Univ. Press, Cambridge.

Shugart, H.H., T.R. Crow & J.M. Hett. 1973. Forest succession models: a rationale and methodology for modeling forest succession over large regions. For. Sci. 19: 202–212.

Usher, M.B. 1966. A matrix approach to the management of renewable resources, with special reference to selection forests. J. Appl. Ecol. 3: 355–367.

Vitousek, P.M. 1977. The regulation of element concentrations in mountain streams in the northeastern United States. Ecol. Monogr. 47: 65–87.

Vitousek, P.M. & W.A. Reiners. 1975. Ecosystem succession and nutrient retention: a hypothesis. Bioscience 25: 376–381.

Whittaker, R.H. 1972. Evolution and measurement of species diversity. Taxon 21: 213–251.

Whittaker, R.H. 1975. Communities and ecosystems. MacMilland, N.Y. 385 pp.

Whittaker, R.H. 1977. Evolution of species diversity in land communities. Evol. Biol. 10: 1–67.

Whittaker, R.H. & P.L. Marks. 1975. Methods of assessing terrestrial productivity. In: H. Lieth & R.H. Whittaker (eds.), Primary productivity of the biosphere. Ecol. Stud. 4: 55–118. Springer-Verlag, N.Y.

Whittaker, R.H. & G.M. Woodwell. 1968. Dimension and production relations of trees and shrubs in the Brookhaven forest, New York. J. Ecol. 56: 1–25.

Yoda, K., T. Kira, H. Ogawa & H. Hozumo. 1963. Self-thinning in overcrowded pure stands under cultivation and natural conditions. J. Biol. Osaka City Univ. 14: 107–129.

Accepted 5 November 1979

# THE INDIVIDUALISTIC NATURE OF PLANT COMMUNITY DEVELOPMENT*

David C. GLENN-LEWIN

Department of Botany, Iowa State University, Ames, Iowa 50011, USA

## Introduction

Most models of plant community development (succession) explicitly or implicitly consider the process to be a series of graded steps culminating in, or converging upon, some relatively stable long-term plant community. Those models which do not proceed by an orderly series of steps nevertheless call for relatively predictable changes in plant communities. Recently, Connell & Slatyer (1977) have seriously questioned the 'facilitation' models of succession, and Olson (1958), Drury & Nisbet (1973), Walker (1970) and Matthews (1979) have questioned the hypothetical convergence of successional pathways. In this paper, I use several sources of evidence to call into question the predictability of plant community change over time. Two of these sources are anecdotal observations, three are reviews of previously published data, and one is research first described here in detail; the quantitative data come from vegetation studies in the north-central United States.

## Review of evidence

### Anecdotal observations

On the grounds of the Chesapeake Bay Center for Environmental Studies, USA, (a branch of the Smithsonian Institution), which is located on the coastal plain of the Bay, are two parcels of land which, according to Center personnel, were abandoned at the same time and under the same conditions. At present, one of these stands is a young elm (*Ulmus*) forest of pole-sized trees with a typical forest undergrowth. The other parcel is vegetated by a

* I thank Jon White, Maryanne Beach, Roger Landers, Arnold van der Valk, Robert Whittaker, Robert Peet, and Mike Chadwick for their ideas and discussions. Jon White participated in the Kalsow Prairie field work.

tangle of poison ivy (*Rhus radicans*) so dense that when traversing the area, one walks across the stems of the plants rather than on the ground. Thus, even though conditions were very similar, plant community development on these two sites has been different.

In the species-rich, old forest of Horseshoe Lake, southern Illinois, USA, (described by Robertson et al 1978), *Sassafras albidum* can be found as large trees sharing the canopy; *S. albidum* is normally thought of as a pioneer species.

### Evidence from published studies

Three recent studies of vegetation dynamics in Iowa, USA, bear directly on the question of plant community dynamics. Cahayla-Wynne & Glenn-Lewin (1978) performed a classification and ordination analysis of the upland forest vegetation on the unglaciated area in extreme northeast Iowa. In those communities that were dominated by white pine (*Pinus strobus*), the most significant component of the shorter trees was white oak (*Quercus alba*). It is the normal expectation (e.g. Anderson & Adams 1978) that pine forests succeed to oak forests, but in the forests of northeast Iowa, increment cores show that the tall, large pines were younger than the oaks. In short, the pine and oak do not show the expected successional relationship.

Second, it has been demonstrated recently that plant community dynamics in prairie glacial marshes of the north-central United States are cyclical, rather than convergent (van der Valk & Davis 1978). The apparent explanation for this behavior is found in the interaction between the 'seed bank' in the marsh substrate and water level fluctuations. Mudflat communities, emergent-species communities, and the lake stage of the marsh are all related to the differential germination properties of the seeds of marsh species in response to periodic drought and flooding of the marsh.

Table 1. Average cover and species richness (number of species) for six coal spoil areas of different ages, according to three classes of substrate quality. Southeast Iowa, USA. From Glenn-Lewin (1978).

| Area | Age | Per Cent Cover | | | Richness | | |
|------|-----|----------|--------------|------|----------|--------------|------|
| | | Non-acid | Intermediate | Acid | Non-acid | Intermediate | Acid |
| Amsberry | 5–7 | 49.8 | 13.3 | 2.7 | 23.7 | 11.0 | 2.5 |
| Fee | 13–16 | 56.0 | 13.6 | 3.7 | 12.0 | 9.4 | 3.2 |
| Harrison | 17 | 34.0 | 17.3 | 0.5 | 14.7 | 7.7 | 0.8 |
| Wilcox | 22–23 | 144.6 | 67.7 | 1.3 | 32.1 | 21.6 | 2.7 |
| Watertor | 26–27 | 77.8 | 53.9 | 10.3 | 27.8 | 20.9 | 7.3 |
| Klein | 36–38 | 148.5 | 42.7 | 6.0 | 23.0 | 15.4 | 3.9 |

Third, in southeastern Iowa are found a large number of coal stripmine spoils ('orphan banks') of varying ages. In an analysis of these orphan banks, Glenn-Lewin (1978) found that, aside from the fact that the trees were larger, there appeared to be no relationship between age of the orphan bank and species composition or community structure (interpreted as species diversity and canopy coverage). Instead, substrate conditions, particularly soil acidity due to the oxidation of pyrites, seemed to be the most significant factor in determining the vegetation of these spoils (Table 1). This pattern was confirmed by reciprocal averaging ordination (Glenn-Lewin 1978). Jonescu (1978) in Saskatchewan, Canada, and Chadwick & Hardiman (1976) have also concluded that the normal models of succession are inappropriate for the vegetation of orphan banks.

### Natural revegetation of a grazed grassland

It is difficult to find in the rich agricultural region of the central United States, areas of natural grassland in which secondary succession is occurring. Fortunately, in 1946, the State of Iowa preserved Kalsow Prairie, a native tall-grass prairie of approximately 65 ha, of which about 13 ha, in the northwest corner, was a heavily grazed pasture. The pasture was preserved at the same time as the prairie, and the fences around the pasture were removed. The pasture was not analyzed when it was abandoned, but probably contained *Poa pratensis*, *Solidago canadensis*, and *Cirsium arvense* as important species, judging from other heavily grazed pastures in the region.

The vegetation of the Kalsow Prairie grazed area was recorded by Brotherson (1969, 1979), who did his field work in 1967. I restudied the same area in 1977. The purposes of these analyses were to determine:

1. In what fashion was revegetation occurring, i.e., what was the pattern of species replacement?
2. What is the rate of revegetation; how fast was the vegetation returning to the state of the adjacent prairie?

### Methods

Brotherson's (1969, 1979) sampling scheme was repeated again in 1977 for comparability between the studies. Thirty plots, each 60 by 78 m, in an arrangement of 5 rows running east and west by 6 rows running north and south, were marked off. Within each of these plots, twenty 0.1 m² (20 × 50 cm) quadrats were placed in a sigmoid fashion and species coverages were estimated in each quadrat. For each plot, coverages were averaged over the 20 small quadrats.

### Results and discussion

Brotherson (1969, 1979) noted that the original fenceline between the grazed area and native prairie was still visible in 1967. That fenceline was no longer visible in 1977. Sixty-five species were recorded in 1967, whereas 87 species were recorded in 1977. The coefficient of community (Sørensen 1948) was 0.73.

Brotherson (1969, 1979) stated that there was no pattern of species diversity in the grazed area in 1967. Fig. 1 demonstrates that in 1977, the plots with the greatest species richness were located adjacent to the native prairie, and those with the fewest numbers of species were located farthest away from the prairie. The concentration of dominance, as calculated by Simpson's (1949) index, demonstrates the converse of the richness pattern; the areas of highest dominance are located away from the prairie (Fig. 2).

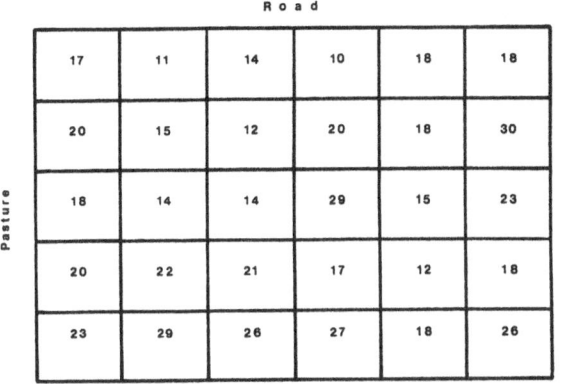

Fig. 1. Species richness in the Kalsow Prairie grazed area, 1977.

Table 2. Constancy and average cover of dominant species on the Kalsow Prairie, Iowa, USA, grazed area, 1977.

| Species | Constancy | Cover |
|---|---|---|
| Poa pratensis | 1.00 | 40.73 |
| Andropogon gerardii | .87 | 39.28 |
| Solidago canadensis | 1.00 | 15.38 |
| Aster ericoides | .97 | 4.10 |
| Sporobolus heterolepis | .27 | 3.68 |
| Bromus inermis | .23 | 2.71 |
| Spartina pectinata | .47 | 2.60 |
| Carex spp. | .80 | 2.50 |
| Helianthus grosseserratus | .43 | 2.17 |

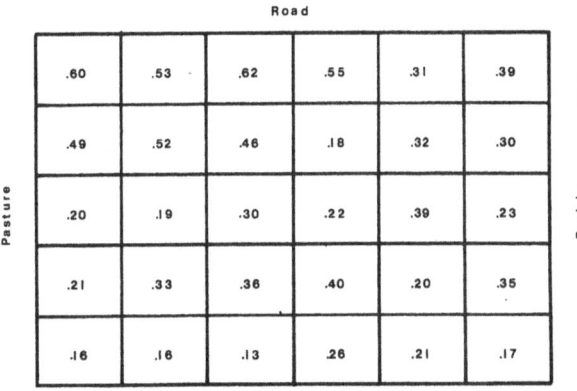

Fig. 2. Concentration of dominance (Simpson's Index) in the Kalsow Prairie grazed area, 1977.

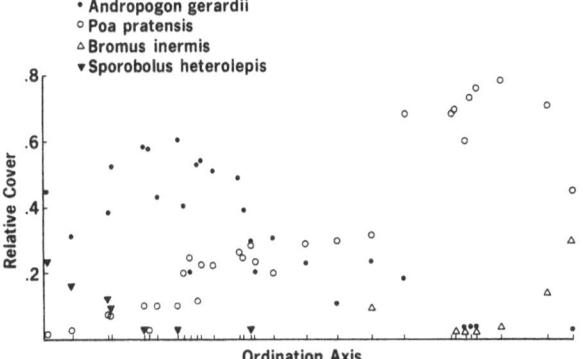

Fig. 3. Reciprocal averaging ordination position of each 60 × 78 m plot in the Kalsow Prairie grazed area, 1977.

On the whole, the Kalsow Prairie pasture was dominated in 1977 by some species which are characteristic of the native prairie, and some which are characteristic of heavily grazed areas (Table 2). However, these species were unevenly distributed over the grazed area. A reciprocal averaging ordination placed the plots nearest the native prairie at one end of the ordination axis, whereas the plots farthest from the prairie fell toward the other end of the axis (Fig. 3). When the four most important grasses were plotted along the reciprocal averaging ordination axis (Fig. 4), the native prairie species *Andropogon gerardii* and *Sporobolus heterolepis* peaked near the end of the ordination axis having the plots nearest the prairie, while *Poa pratensis* and *Bromus inermis*, species increasing with grazing, had peak abundances at the other end of the ordination axis, which corresponded to the portion of the

- Andropogon gerardii
○ Poa pratensis
△ Bromus inermis
▼ Sporobolus heterolepis

**Relative cover of four important grass species along the reciprocal averaging ordination axis, Kalsow Prairie grazed area, 1977.**

Fig. 4. Relative cover of four important grass species along the reciprocal averaging ordination axis, Kalsow Prairie grazed area, 1977.

143

grazed area farthest away from the native prairie. By Spearman rank correlation (Steel & Torrie 1960), ordination position near the native prairie was related to higher species richness ($r_s = 0.65$, $p < 0.001$).

Brotherson (1969, 1979) plotted the abundance of *Andropogon gerardii* throughout the grazed area for 1967 (Fig. 5). A similar plot for 1977 is shown in Fig. 6. Brotherson (1969, 1979) was of the opinion that *A. gerardii* had formed an invasion front in 1967. A comparison with 1977, however, indicates *A. gerardii* did not form such a front, since the apparent front had not moved forward. Instead what had happened is that the relative abundance of *A. gerardii* had begun to differentiate from place to place.

It is clear from the facts that: 1) the fenceline was no longer visible, 2) patterns of species diversity had developed, 3) an ordination analysis produced patterns of vegetation related to the geography of the grazed area, and 4) the differentiation of the *Andropogon gerardii* population, that vegetational changes have occurred in the Kalsow Prairie grazed area between 1967 and 1977. However, the changes that have occurred have all done so mainly in those plots adjacent to the native prairie; the farther away from the native prairie, the less change has occurred in the grazed area. Only along the prairie edge were species found such as *Liatris asper*, *Sporobolus heterolepis*, *Aster laevis*, *Gentiana puberula*, *Phlox pilosa*, all species characteristic of the native prairie.

Rather than changing as a series of different species populations, the vegetation in the Kalsow Prairie grazed area changed by a process of the native prairie encroaching

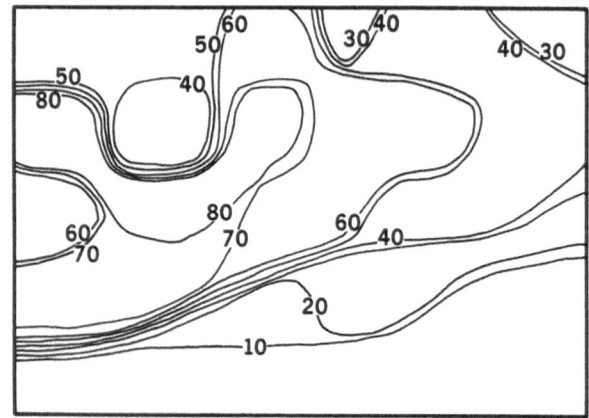

**Distribution of <u>Andropogon gerardii</u>, 1977.**

Fig. 6. The distribution of per cent cover of *Andropogon gerardii* in 1977.

from the edge as a more-or-less whole prairie vegetation, and it seemed most likely that this encroachment occurred by vegetative reproduction. If this is true, than the rate of revegetation is simply a matter of comparing the total distance to the edge of the grazed area with the distance that the vegetation has moved over the last approximately 30 years. Inspection in the field shows that the majority of the native prairie species have moved perhaps 10 to 20 m since the prairie was preserved. Since, in order to totally revegetate the pasture, the native prairie vegetation must move about 300 m, it appears that it will take roughly 450 to 900 yr for the process to be completed. Such a long time span may be an overestimate, but even if so, replacement of the grazed area species by native prairie species will take a long time.

This analysis of the dynamics of vegetation replacement in a north-central U.S. grazed grassland leads to three conclusions:

1. Revegetation of the grazed area is by encroachment of a whole prairie vegetation from the edge, rather than a series of species replacements as predicted by classical successional models.

2. If 1, above, is accurate, then it is unlikely that abandoned areas will ever get to a native prairie stage unless such an abandoned area is adjacent to a native prairie, or unless the area is intentionally planted.

3. The rate of revegetation in disturbed grasslands is very slow.

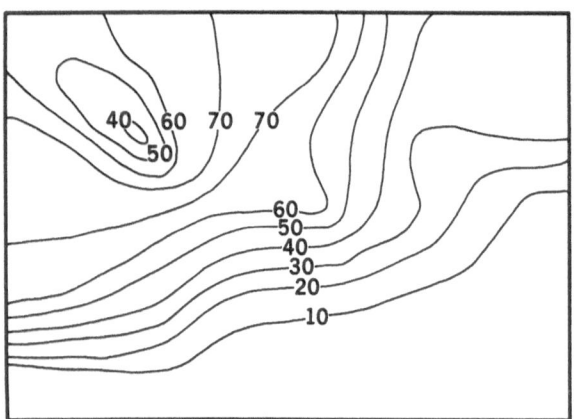

**Percent cover of <u>Andropogon gerardii</u>, 1967. From Brotherson (1968).**

Fig. 5. The distribution of per cent of *Andropogon gerardii*, 1967. From Brotherson (1969).

144

## Concluding remarks

The evidence brought to bear on the question of plant community development in this paper comes from simple observation, from studies that compared several sites of differing ages, from studies that examined a site over time, and from citations of others. The evidence indicates that:

1. Species may behave in an anamolous manner, i.e., species may take on different roles in community development than those usually described for them.
2. Vegetation development may be divergent under very similar environmental conditions.
3. Vegetation development may show little or no relationship to time when an environmental factor(s) is very important.
4. Vegetation development may occur by other means than species replacements via seed populations. Such change is very slow.

In the grassland region of Nort America, at least, there may be no such thing as succession in its usual meaning. Rather, plant community development appears to be individualistic in character. Such individualistic plant community development is site-specific and relatively unpredictable, i.e., with a significant random component, at least in terms of species composition.

If this is true, then ecologists perhaps should begin to look in different directions for a better understanding of plant community dynamics. Clearly, vegetation dynamics are related to seed dynamics. In addition, plant life histories are important, and the examples of the orphan banks illustrate the importance of microsite environmental variation. Research in these areas may shed new light on the process of vegetation change.

## Summary

A body of anecdotal and experimental evidence raises questions about the common model of plant community succession, in which each temporal vegetation stage facilitates the development of the next stage. Several lines of evidence are discussed which emphasize the individualistic, independent nature of plant community development:

1) There are a number of instances of anamolous behaviour of 'pioneer' or 'early successional' species;
2) In shallow marshes, vegetational change is the result of the interaction between the stored 'seed bank' of the the ecosystem and the fluctuating water regime;
3) On abandoned coal spoils, vegetational composition and structure are almost entirely accounted for by substrate conditions rather than age differences of the spoils;
4) In overgrazed natural grasslands, adjacent to ungrazed areas, revegetation is primarily by vegetative reproduction of the native species creeping in from the edge as a whole vegetation complex, rather than seeding throughout in stages.

Therefore, plant succession can be interpreted as an individualistic result of the interaction of disturbance (both kind and severity) and the predominant reproductive life histories of the plant species.

It is suggested that future research efforts to understand vegetational change should be directed to the interactions between seed dynamics, asexual/sexual life histories, and microsite environmental variation.

## References

Anderson, R.C. & D.E. Adams. 1978. Species replacement patterns in central Illinois white oak forests. In: P.E. Pope (ed.), Cent. Hardwood Forest Conf. II Proc. p. 284–301.

Brotherson, J.D. 1969. Species composition, distribution, and phytosociology of Kalsow Prairie, a mesic tall-grass prairie in Iowa. Ph. D. dissertation, Iowa State Univ. 196 pp.

Brotherson, J.D. & R.Q. Landers. 1979. Recovery from severe grazing in an Iowa tall-grass prairie. In: D.C. Glenn-Lewin and R.Q. Landers (eds.), Proc. Fifth Midwest Prairie Conf. p. 51–56.

Cahayla-Wynne, R. & D.C. Glenn-Lewin. 1978. The forest vegetation of the Driftless area, Northeast Iowa. Amer. Midl. Nat. 100: 307–319.

Chadwick, M.J. & K.M. Hardiman. 1976. Vegetating colliery spoil. In: Papers of the Land Reclamation Conference, pp. 421–42. Grays, Thurrock Borough Council.

Connell, J.H. & R.O. Slatyer. 1977. Mechanisms of succession in natural communities and their role in community stability and organization. Amer. Nat. 111: 1119–1144.

Drury, W.H. & I.C.T. Nisbit. 1973. Succession. J. Arnold Arboretum Harvard Univ. 54: 331–368.

Glenn-Lewin, D.C. 1978. Natural revegetation of acid coal spoils in southeast Iowa. In: M.K. Wali (ed.), Proc. 1st Internat. Cong. Energy and the Ecosystem. (in press).

Jonescu, M.E. 1978. Natural revegetation of strip mined land in the lignite coal fields of southeastern Saskatchewan. In: M.K. Wali (ed.), Proc. 1st Internat. Cong. Energy and the Ecosystem. (in press).

Matthews, J.A. 1979. A study of the variability of some successional and climax plant assemblage-types using multiple discriminant analysis. J. Ecol. 67: 255–272.

Olson, J.S. 1958. Rates of succession and soil changes on southern Lake Michigan sand dunes. Bot. Gazette 119: 125–170.

Robertson, P.A., G.T. Weaver & J.A. Cavanaugh. 1978. Vegetation and tree species patterns near the northern terminus of the southern floodplain forest. Ecol. Monogr. 48: 249–267.

Simpson, E.H. 1949. Measurement of diversity. Nature 163: 688.

Sørensen, T.A. 1948. A method of establishing groups of equal amplitude in plant sociology based on similarity of species content, and its application to analyses of the vegetation on Danish commons. K. Danske Vidensk. Selsk. Biol. Skr. 5: 1–34.

Steel, R.G.D. & J.H. Torrie. 1960. Principles and procedures of statistics. McGraw-Hill, N.Y. 481 pp.

Valk, A. van der & C.B. Davis. 1978. The role of seed banks in the vegetation dynamics of prairie glacial marshes. Ecology 59: 322–335.

Walker, D. 1970. Direction and rate in some British post-glacial hydroseres. In: D. Walker & R.G. West (eds.), Studies in the vegetational history of the British Isles, Cambridge Univ. Press, London. p. 117–139.

Accepted 5 November 1979

# VEGETATION DYNAMICS OR ECOSYSTEM DYNAMICS: DYNAMIC SUFFICIENCY IN SUCCESSION THEORY

Robert VAN HULST

Department of Biology, Bishop's University, Lennoxville, Quebec, Canada J1M 1Z7

Keywords:
Dynamics, Ecosystem, Markovian model, Succession, Vegetation

## Introduction

Recently several authors have claimed that long term vegetation change, at least in the case of forest ecosystems, can usefully and quite accurately be described by stationary Markov chain models (Anderson, 1966; Waggoner & Stephens, 1970; Horn, 1974, 1975a, 1975b, 1976) or their deterministic counterpart, coupled differential equation models (Shugart et al., 1973). Markov chain models do indeed form a class of unusually versatile and well-studied models and their exploratory use as models of succession would seem quite appropriate (for an introduction see van Hulst 1979 and references therein). Markov models, for example, can mimic not only simple linear succession (with species or community B replacing species or community A  C replacing B, and so on), but also successions involving such phenomena as reversals or 'sticky' states, cyclical successions, indeterminate situations, a gradual approach to a steady state, and several other complications (see e.g. Horn 1976).

However, there is one state of affairs which is excluded by the Markov model (and, for that matter, by the differential equation model): future states of the system cannot depend on the system's past states, only on its present state. In other words: how the system reached its present state is irrelevant as far as the prediction of future states is concerned. Whether this assumption, known as the Markovity assumption, is justified or not for an actual data set can be tested statistically (see Billingsley 1961; van Hulst 1979 for the appropriate tests) and, while it is the only crucial assumption of the model, it would also seem an assumption which, on biological grounds, is unlikely to be satisfied. Yet none of the recent applications of Markov chains to succession theory include a direct test of the

Markovity assumption. Instead, various forms of indirect evidence are provided for the appropriateness of the model. Waggoner & Stephens (1970) compare (visually) a four-step transition probability matrix estimated from their raw data with one generated from the one-step transition probability matrix by raising it to the 4th power. They also inspect the steady-state configuration generated by their model and conclude that it is more-or-less realistic. Horn (1975a, b) similarly compares the predicted steady state of his forest with the actual present forest composition and is satisfied that his model is reasonably accurate. Bellefleur (1979) compares model predictions and reality using the Kolmogorov-Smirnov statistic.

Several serious problems arise, however, in these indirect 'tests' of the Markovity assumption. First, the practice often degenerates into curve-fitting, as the same data used to estimate the model parameters are then used to 'validate' the model. With the large number of parameters to be fitted one can probably reproduce almost any kind of data.

Second, even if separate data sets would have been used (note that none of the authors cited above did!) it would seem that these indirect 'tests' are not very robust: computer simulation suggests that it is relatively easy to obtain an 'approximate fit', especially if one does not run the model until equilibrium is reached (because it is only after many iterations that discrepancies start to really show up).

Third, it can then be argued – and with some justification – that persistent perturbation will in practice prevent the systems from reaching the theoretical steady state anyway. By introducing a further 'disturbance' matrix into the model (Horn 1976) an additional array of constants is made available to the ecologist with which the model behavior can be brought in correspondence with reality (one is temped to say: *any* reality).

Admittedly, direct tests of the Markovity assumption are difficult in that they require many data points. They are not impossible, however, and this avenue should be explored much more thoroughly. But there is also another road: we may ask ourselves under what *biological* assumptions we might expect the Markovity assumption to hold. It is this road which I wish to explore here. First, however, a digression into some formal matters, relating in particular to the nature of a process description, seems appropriate.

## Dynamic sufficiency

Most mathematical process descriptions (such as the laws of motion) are constructed according to the same recipe: one characteristically separates *state variables*, which enter into the dynamic model (or 'law'), from *initial conditions*, about which no pronouncements are made. In succession theory the state variables are usually vegetation data: either some measure of the abundance of the various species present, or community identity (e.g. Anderson 1966, Horn 1975a, b, 1976, Bellefleur 1979). Occasionally age structure (e.g. Usher 1966) or other ecosystem components (e.g. Holling et al. 1975) are included.

It is important to note here that description and theory building are interdependent processes. In describing reality we single out certain aspects in which we hope to demonstrate some regular predictable behaviour. The regularity and predictability of these aspects of reality is mimicked by the regularity and predictability in our theoretical constructs, which are usually patterned on some branch of mathematics. The unlawful, irregular aspects of reality are relegated to the initial conditions. But not any description will be compatible with an effective (useful, interesting, elegant) theory, just as not any theory will correspond to a feasible way of describing reality. In particular, a description may be dynamically insufficient, i.e. it may not allow in conjunction with the theoretical framework used, for predicting future systems behaviour (see Lewontin 1974, for a lucid discussion of these matters). One manifestation of dynamic insufficiency in mathematical models is non-Markovity: the future states of the system depend not only on the present state but also on the particular way in which the system arrived at the present state. Non-Markovity is therefore not a property of the world (or some as-

pect of it), it is a property of a particular way of interpreting reality (consequently there is little sense in asking whether succession is Markovian or not). In the context of succession theory, then, we need to ask whether vegetation data by itself forms a dynamically sufficient description of a plant community, enabling us to predict its future fate with some accuracy. It would seem that several arguments militate against this possibility.

When, indeed, would vegetation data alone be sufficient to predict a plant community's future states? The answer, clearly, is: if and only if, there is no 'memory' in the system for past states. This memory could take a number of different forms. First, it could reside in the environment of the community in question: in the soil, in the structure of herbivore or pathogen populations. And second, it could be a factor *internal* to the community: in the age structure of some populations, in spatial imbalances such as locally density dependent recruitment (Horn 1976), etc. In both cases mentioned the past course of vegetation change and past disturbances may leave traces which continue to influence future vegetation change for some time. The system, in other words, carries with it a memory of past events, a phenomenon one might christen biological hysteresis. The only condition under which this would not be the case is the biologically unlikely one in which differences in environmental states are immediately and uniquely expressed by differences in vegetation. For in this case the only relevant state variables are the vegetational ones. Whereas this condition is perhaps approximately fulfilled in the case of changes in the light regime at a site, there remain many environmental factors, which will not provide immediate reactions on the part of the vegetation. In any case, it is now generally assumed that the distribution and abundance of plants is not directly dependent on physical-chemical habitat conditions (e.g. Walter 1973) and the condition seems therefore unlikely to hold. Indeed many examples of biological hysteresis can be found in the case of forest communities, to which I will limit myself in what follows.

## Hysteresis in forest communities

One way in which past events may influence future vegetation change – in forest communities possibly the most important one – is through the intermediary of the forest soil. Indeed, the classical explanation of successional change, with early successional species preparing the environment for later successional ones, involves hysteresis effects, presumably through changes in such soil factors as organic matter content and water regime. Evidence for the importance of these effects is mainly anecdotal and in several studied cases secular changes in the water table, rather than community-caused 'reaction' (Clements 1916) or 'facilitation' (Horn 1976) proved to be responsible for community change (Olson 1958, Heinselman 1963, Walker 1970). Nevertheless, there are also well documented examples of soil-mediated hysteresis effects. These include the fertility improving effect of nitrifying bacteria in root nodules of alders (Crocker & Major 1955); improved nutrient circulation due to the activity of forest trees (Armson 1977); the formation of podzols under certain vegetation types (Spurr & Barnes 1973, Armson 1977, Damman, 1971); soil degradation as a result of previous agricultural use (Stone & Leaf 1967) and the effects of fire on the soil (Armson 1977). The nature of the humus layer and of the soil biota are obvious further candidates for originating hysteresis effects, and in actual practice the long-term effects of previous vegetation often include soil chemistry and physical structure as well as its biological make-up (Spurr & Barnes 1973).

In addition to soil mediated effects purely biological effects can influence a forest's future fate. Examples which are particularly well-studied include such herbivores as the spruce-budworm in North-America (Morris 1963; Baskerville 1971) or the persistence of pathogens (Spurr & Barnes 1973).

Finally, two potentially important sources of hysteresis in forest communities involve the spatial distribution and the age structure of the populations involved. Recruitment in forest tree species may be dependent on local density (Horn 1976), in which case spatial structure influences community composition. In its age structure a population may similarly retain a memory of past events (disturbances, or past abundance). If, in addition, age structure influences future recruitment (as seems very likely), then age structure is another possible source of hysteresis. The one age class which is often of great importance is that of the (buried) seeds. The important role of pin cherry in North American forest succession, is a well known example of this (Marks 1974). Many other examples of age structure effects on population dynamics are reported in Harper (1977); see also van Hulst (1978).

### Equilibrium versus non-equilibrium dynamics

In view of the apparently frequent occurrence of hysteresis in forest communities, one may expect non-Markovity to be a common condition in forest dynamics models. Somewhat circumstantial evidence for this comes from an attempt to model secondary forest succession in British Columbia with Markov chain models (Bellefleur 1979). Here it was found that the assumptions of the model, and in particular the Markovity assumption, did not hold (the assumption, however, was not tested directly). Similarly, Hahn & Leary (1974) claim that the differential equation model employed by Shugart et al. (1973) is insufficient and unlikely to shed any light on the reality of forest succession.

If a vegetation description in the context of a Markov chain (or differential equation) model of vegetation dynamics is indeed dynamically insufficient the situation can be remedied, at least in theory, by simply extending the state space. We may have to include soil and humus characteristics as state variables, and perhaps information concerning other ecosystems components (herbivores, pathogens) as well. This would complicate the way in which we describe an ecosystem, while maintaining a simple theoretical framework.

Another way to proceed, obviously, is to develop a different theoretical framework, preferably one which is not (as Markov-chains and differential equation models are) equilibrium-oriented. My contention here is that Markov chains are an obvious mathematical analogue of a more or less classical succession which slowly approaches a climax state. Recently, however, several authors have claimed that such a steady state endpoint of succession is, in fact, rarely attained (Henry & Swan 1974, Connell & Slatyer 1977, Connell 1978, Hubbell 1979). One might, with Horn (1976) attempt to save the Markov model by introducing 'disturbance' matrices, but unless one has an explicit recipe on how to prepare such a matrix (and Horn 1976, does not provide one), it is hard to see what is gained with such an approach. In fact, two specific problems arise if Markov chain models are to be used in situations far from equilibrium. The first difficulty arises in the context of parameter estimation, which is most readily and most precisely achieved at or near equilibrium. If the equilibrium is in fact never attained because of random disturbances, then it will become very difficult to estimate model parameters. Related to this, validation of a Markovian succession model should preferably be attempted on the basis of the steady state produced by the model, since it is here that discrepancies show up most clearly. This again presupposes that there is a corresponding steady state in reality.

A second and more fundamental difficulty involves the separation of 'normal system behaviour' and disturbance. Unless both are known in considerable detail it will be impossible to make a clear separation. The problem here seems to be a very general one: in ecology it is often difficult to shield the system of interest from external perturbations. In fact, the external perturbations seem to determine to a large extent the system's behaviour! Connell (1978) discusses a number of hypotheses based on a non-equilibrium species dynamics (with species composition being seldom in equilibrium), and attempts to explain the relationship between diversity and disturbance. Hubbell (1979) presents a mathematical version of Connell's (1978) argument, relating relative species abundance patterns to a disturbance gradient. A similar model is applied by van Hulst (in prep.) to hardwood forests in Eastern North America. Models such as these may be more promising in the study of diversity patterns than in the actual prediction of future vegetation change. However, as long as our understanding of the role of such diversity patterns, and of the natural disturbances by which they seem to be maintained, is as limited as it is, little progress seems possible in the more applied aspects of ecosystem dynamics.

## Conclusion

The adequacy of Markovian vegetation models cannot be judged solely by their ability to generate reasonable predictions after large numbers of parameters have been fitted. However, statistical tests are known to test the Markov property directly, although the required data are difficult to obtain. Indirect evidence suggests that for a number of forest successions the assumption is probably not satisfied.

On biological grounds this is not unreasonable: many examples of lasting effects of past vegetation are reported in the literature, especially in the forestry literature. Some of these have been reviewed here. The effect of such 'hysteresis' phenomena would be to make a description of a forest community in terms of vegetation data alone dynamically insufficient. This in turn leads to a violation of the Markovity assumption in Markovian succession models, for the only condition under which Markovity would be retained, namely if differences in environmental states are immediately and uniquely expressed in vegetation differences, is unlikely to be satisfied for many environmental factors.

Two quite different ways to deal with this problem seem to be open to us. The first is to retain the Markovian model and to extend the state description: instead of vegetation dynamics we should study ecosystem dynamics.

The second approach throws out the model, which, being essentially an equilibrium model, is judged inappropriate anyway, and focuses on the direct effects of disturbances. Both seem to promise a better understanding, if not a solution, of an extraordinary difficult problem: the relation between continuity and change in natural communities.

## Summary

Most mathematical models of processes are based on a separation between state variables and initial conditions. If the state variables only include the present states of the system, not its past states, then the model is said to be Markovian and the full set of the state variables provides a dynamically complete description of the process.

Recently a number of claims have been made to the effect that forest succession can be regarded as a Markovian process with vegetational composition as the state variables. On theoretical grounds such a description would be expected satisfactory if, and only if, differences in environmental states are immediately and uniquely expressed by differences in vegetation. This will be approximately true if light competition is the sole driving force of succession, but in the case of successional change driven by changes in other ecosystem components (such as soil moisture and soil-nutrient status) this condition will not in general be fulfilled.

The adequacy of Markovian vegetation models cannot be judged solely by their ability to generate reasonable predictions after large numbers of parameters have been fitted. However, there exist statistical tests to test the Markov property directly, although the required data are hard to obtain. In at least two cases reported in the forestry literature present vegetational composition alone was judged not a sufficient basis for the prediction of future vegetational change. Apparently the introduction of other variables is necessary to make the state space complete. The conclusion therefore emerges that students of vegetational succession should either leave the narrow domain of vegetation dynamics in order to study ecosystem dynamics, or explore different succession models. The use of non-equilibrium models of vegetation dynamics is briefly discussed here.

## References

Anderson, M.C. 1966. Ecological groupings of plants. Nature 212: 54–56.

Armson, K.A. 1977. Forest Soils: Properties and Processes. University of Toronto Press, Toronto, 390 pp.

Baskerville, G.L. 1971. The fir-spruce-birch forest and the budworm. Report Forestry Service, Canada Dept. Envir., Fredericton, N.B., 111 pp.

Bellefleur, P. 1979. Markov models of forest-type secondary succession in British Columbia. Can. J. Forest Res. (in press).

Billingsley, P. 1961. Statistical methods in Markov chains. Ann. Math. Stat. 32: 12–140.

Clements, F.E. 1916. Plant succession. Carnegie Inst. Wash. Publ. 242, 512 pp.

Connell, J.H. 1978. Diversity in tropical rain forests and coral reefs. Science 199: 1302–1310.

Connell, J.H. & R.O. Slatyer. 1977. Mechanisms of succession in natural communities and their role in community stability and organization. Amer. Nat. 111: 1119–1144.

Crocker, R.L. & J. Major. 1955. Soil development in relation to vegetation and surface age at Glacier Bay, Alaska. J. Ecol. 42: 427–448.

Damman, A.W.H. 1971. Effect of vegetation changes on the fertility of a Newfoundland forest site. Ecol. Monogr. 41: 253–270.

Hahn, J.T. & R.A. Leary, 1974. Test of a model of forest succession. Forest Sci. 20: 212.

Harper, J.L. 1977. Population biology of plants. Academic Press, London, 892 pp.

Heinselman, M.L. 1963. Forest sites, bog processes and peatland types in the glacial Lake Agassiz region, Minnesota, Ecol. Monogr. 33: 327–374.

Henry, I.D. & T.M.A. Swan. 1974. Reconstructing forest history from live and dead plant material – an approach to the study of forest succession in southwest New Hampshire. Ecology 55: 772–783.

Holling, C.S., G.B. Dantzig, G. Baskerville, D.D. Jones & W.C. Clark, 1975. A Case study of forest ecosystem/pest management. Inst. Res. Ecol. Univ. British Columbia, 42 pp.

Horn, H.S. 1974. The ecology of secondary succession. Ann. Rev. Ecol. Syst. 5: 25–37.

Horn, H.S. 1975a. Forest succession. Scient. Amer. 232: 90–98.

Horn, H.S. 1975b. Markovian properties of forest succession. In: M.L. Cody & J.M. Diamond (eds.), Ecology and evolution of communities. p. 196–211. Harvard University Press, Cambridge.

Horn, H.S. 1976. Succession. In: R.M. May (ed.), Theoretical ecology. p. 187–204. W.B. Saunders, Philadelphia, 317 pp.

Hubbell, S.P. 1979. Tree dispersion, abundance and diversity in a tropical dry forest. Science 203: 1299–1309.

Hulst, R. van. 1978. The dynamics of vegetation: patterns of environmental and vegetational change. Vegetatio 38: 65–75.

Hulst, R. van 1979. On the dynamics of vegetation: Markov chains as models of succession. Vegetatio 40: 111–111.

Lewontin, R.C. 1974. The genetic basis of evolutionary change. Columbia Univ. Press. 346 pp.

Marks, P.L. 1974. The Role of pin cherry (Prunus pensylvanica L.) in the maintenance of stability in northern hardwood ecosystems. Ecol. Monogr. 44: 73–88.

Morris, R.H. 1963. The dynamics of epidemic spruce budworm populations. Mem. Entomol. Soc. Can. 31: 1–322.

Olson, J.S. 1958. Rates of succession and soil changes on Southern Lake Michigan sand dunes. Bot. Gazette 119: 125–170.

Shugart, H.H. Jr., T.R. Crow & J.M. Hett. 1973. Forest succession models: a rationale and methodology for modelling forest succession over large regions. Forest Sci. 19: 203–212.

Spurr, S.H. & Barnes, B.V. 1973. Forest Ecology. 2nd ed. Ronald Press, New York, 571 pp.

Stone, E.L. & A.L. Leaf. 1967. Potassium deficiency and response in young conifer forests in eastern North America. In: Proc. coll. forest fertilization, Finland 1967, p. 217–219. Internat. Potash Inst., Berne.

Usher, M.B. 1966. A matrix approach to the management of renewable resources with special reference to selection forests. J. Appl. Ecol. 3: 355–367.

Waggoner, P.E. & G.R. Stephens. 1970. Transition probabilities for a forest. Nature 225: 1160–1161.

Walker, D. 1970. Direction and rate in some British postglacial hydroseres. In: D. Walker & R. West (eds.), The vegetational history of the British Isles, p. 117–139. Cambridge University Press, Cambridge.

Walter, H. 1973. Vegetation of the earth. Springer, New York, 237 pp. (originally: Vegetationszonen und Klima, Verlag Eugen Ulmer, Stuttgart, 2nd ed. 1973).

Accepted 5 November 1979